Planets
Beyond

The Wiley Science Editions

Planets Beyond

Discovering the Outer Solar System

Mark Littmann

WILEY

Wiley Science Editions

John Wiley & Sons, Inc.

New York • Chichester • Brisbane • Toronto • Singapore

Publisher: Stephen Kippur
Editor: David Sobel
Managing Editor: Andrew Hoffer
Assistant Managing Editor: Corinne McCormick
Editing, Design, & Production: G&H SOHO, Ltd.
Illustrator: Charles Messing

Library of Congress Cataloging-in-Publication Data

Littmann, Mark. 1939—
    Planets beyond: discovering the outer solar system / Mark Littmann.
        p.   cm.   (Wiley science editions)
    Bibliography : p. 271
    ISBN 0-471-61128-X
    1. Uranus (Planet) 2. Neptune (Planet) 3. Pluto (Planet)
I. Title. II. Series.
QB681.L58 1988   88-20498
523.4'7—dc19   CIP

Printed in the United States of America

88 89 10 9 8 7 6 5 4 3 2 1

*For Peggy
with love*

Preface

I hope you like this story. The discoveries of Uranus, Neptune, and Pluto were among the first sagas I encountered when I began to read about astronomy. They stuck fast to my sense of drama and irony and I wanted someday to retell them, ideally with the power with which they struck me.

Now might be an appropriate time to try. NASA's *Voyager 2* spacecraft, having succeeded magnificently at Jupiter, Saturn, and Uranus, is about to complete its 12-year Grand Tour of the giant outer planets as it encounters Neptune in August 1989. The great legacy of *Voyager 2* data from Uranus is only now reaching full evaluation.

And although no spacecraft is planned to visit Pluto, Pluto has come part way to visit us. From 1979 to 1999 it is closer to the Sun than Neptune. In 1989 it is closer to us than at any other time in its 248-year orbit. At the same time, Pluto and its moon Charon are engaged in a rare frenzy of mutual eclipses. Pluto's closeness and eclipses are allowing astronomers to extract from the ninth planet more information than ever before possible.

It used to be that Uranus, Neptune, and Pluto each received a vague paragraph in astronomy textbooks. Not much of interest. Not much was known. Now each is a world surrounded by worlds with unique and fascinating stories to tell us about their evolution and, because they have been less changed by the Sun, about the formation of the solar system itself. These geologic worlds have stories as worthy of recounting as their discoveries.

Thus, in part, this book is a progress report—a way to know what is known and suspected so that we can better appreciate what is being discovered.

Here is a history of long-ago events; here are modern events that are shaping history.

In these pages are striking personalities and high drama; profound confirmations of universal scientific laws; human arrogance and stupidity; intellectual courage and perseverance; triumph, humor, irony, and tragedy. Here are scientific discoveries as they were made and modern scientific research as it bounds and staggers forward with uncertainty and brilliance.

Acknowledgments

Spend a moment here, please.

My name is on the cover, but this book would not have happened or had what quality you find in it without some very special and generous people.

It is an extensive list because wherever I turned in need, a leading professional was graciously willing to provide me with information, ideas, and encouragement. These experts, despite their busy schedules, went far beyond what I ever would have thought possible or proper to request. Previously they had my admiration as scholars. Now they also have my admiration as human beings.

They saved me from many a misstatement and flawed explanation. They even saved me on occasion from my writing style.

The errors that remain in this book are mine. (Let me know what you find; I hope to have the chance to fix them.)

When in this book you find the names of people never fully identified anywhere else, when you find the phrasing precise, you know you have encountered the touch of Ruth S. Freitag of the Library of Congress. She has at her fingertips the most amazing information. It was she who contributed the vignette on the automatic asteroid finder. Her encyclopedic knowledge and unfailing good spirits make me very fortunate indeed to have her as a friend and mentor. She and Don Yeomans are currently writing a wonderful book on comets, to be published by Wiley.

Two noted scientists (and authors) read my entire manuscript and offered extremely helpful suggestions for improvements together with warm encouragement. My sincere thanks to Dr. Thomas R. McDonough, California Institute of Technology, and Dr. Andrew Szentgyorgyi, Columbia University.

Several distinguished scientists and engineers who made major discoveries or who work at the heart of major projects have contributed vignettes to *Planets Beyond*. By their graciousness, these researchers have provided us with special insight into the progress of science and technology. They are:

Dr. James W. Christy, Hughes Aircraft Corporation
Dr. Gary A. Flandro, Georgia Institute of Technology
William J. O'Neil, Jet Propulsion Laboratory
Rex Ridenoure, Jet Propulsion Laboratory
Dr. E. Myles Standish, Jr., Jet Propulsion Laboratory
Professor Clyde W. Tombaugh, New Mexico State University
Dr. Donald K. Yeomans, Jet Propulsion Laboratory.

Drs. Christy, Flandro, Standish, Tombaugh, and Yeomans also helped me greatly with portions of the manuscript in which I tried to tell of their discoveries and research.

A number of very generous scientists and engineers found time to read and provide very significant improvements for one or several chapters. To them, too, I am very grateful:

Dr. Robert Hamilton Brown, Jet Propulsion Laboratory
Dr. Marc W. Buie, University of Hawaii
Dr. Robert J. Cesarone, Jet Propulsion Laboratory
Dr. Dale P. Cruikshank, NASA Ames Research Center
Candace J. Hansen, Jet Propulsion Laboratory
Dr. Charles Kohlhase, Jet Propulsion Laboratory
Dr. William B. McKinnon, Washington University
Dr. Mark V. Sykes, University of Arizona.

Planets Beyond could not have included a chapter on the search for a tenth planet without the enthusiastic explanations and candid advice of the researchers in this field. I sincerely appreciate their help:

Dr. John Anderson, Jet Propulsion Laboratory
Dr. Thomas J. Chester and Michael Melnyk, Jet Propulsion Laboratory
Dr. Robert S. Harrington, U.S. Naval Observatory
Charles T. Kowal, Space Telescope Science Institute
Dr. Conley Powell, Teledyne-Brown Engineering
Dr. Thomas C. Van Flandern, VF Associates
Dr. Daniel P. Whitmire and Dr. John J. Matese, University of
   Southwestern Louisiana.

Often, I had questions to ask—many questions—about articles that I read, about differing interpretations, about ways to explain difficult concepts. What a privilege it was to be able to call on experts. They were terrific:

Dr. Edward Bowell, Lowell Observatory
Dr. Donald L. Gray, Jet Propulsion Laboratory
Dr. William B. Hubbard, University of Arizona

Dr. Andrew P. Ingersoll, California Institute of Technology
Dr. Stanley L. Jaki, Seton Hall University
Dr. William S. Kurth, University of Iowa
Dr. Jack J. Lissauer, State University of New York at Stony Brook
Dr. Brian Marsden, Smithsonian Astrophysical Observatory
Dr. Ellis D. Miner, Jet Propulsion Laboratory
Dr. Philip D. Nicholson, Cornell University
Dr. P. Kenneth Seidelmann, U.S. Naval Observatory
Dr. Eugene M. Shoemaker, U.S. Geological Survey
Dr. David J. Stevenson, California Institute of Technology
Dr. David J. Tholen, University of Hawaii
Dr. Laurence M. Trafton, University of Texas
Dr. Scott Tremaine, University of Toronto
Dr. Jack Wisdom, Massachusetts Institute of Technology.

Many people went far out of their way to help me with information,
illustrations, and special resources for this book. I am very grateful to:

Dr. Roy M. Batson, U.S. Geological Survey
Richard Berry, *Astronomy* magazine
Gail Cleere, U.S. Naval Observatory
Brenda Corbin, U.S. Naval Observatory
James A. DeYoung, U.S. Naval Observatory
Dr. Owen Gingerich, Harvard University
Dr. Dieter B. Herrmann, Archenhold Observatory, German Democratic
    Republic
Thomas Jaqua and Althea Washington, NASA
Ruth Leerhoff, San Diego State University Library
Tony Moller, U.S. Naval Observatory
Adam J. Perkins, Royal Greenwich Observatory
Venetia Burney (Mrs. E. Maxwell) Phair of Epsom, England
Jurrie van der Woude, Jet Propulsion Laboratory
James W. Young, Table Mountain Observatory, California Institute of
    Technology.

To the staffs of the Loyola College, Towson State University, and
Baltimore County Libraries, you were wonderful and considerate and
efficient. I wish I could thank you all by name here.
    The people listed above helped to get this work down on paper. Then
the professionals at John Wiley & Sons and G&H SOHO turned it into
a book. You are reading *Planets Beyond* because of the dedication and
skill of Andrew Hoffer, Ruth Greif, and Corinne McCormick at Wiley
and David Sassian and Claire McKean at G&H SOHO.
    A very special separate note of thanks to David Sobel, editor of this

work and of all the Wiley Science Editions, for his enthusiasm and confidence and wise suggestions.

Finally, there are certain generous and talented souls who contributed in unique and deeply appreciated ways:

Muriel Littmann, my mother
Lewis Littmann, my father, who saw this project started and whom we all miss deeply
Beth Littmann, my daughter
David and Esther Littmann
Carl Littmann
Ann and Paul Rappoport
Jane Littmann
Bea and Tom Owens
Anita and David Carstens
Sarah Althoff
Frank Bigger, American Chemical Society

and most especially, my wife, Peggy, to whom this book is lovingly dedicated.

Contents

Planets Beyond

CHAPTER 1

The Discovery of Uranus

". . . it was that night its turn to be discovered."
William Herschel describing his discovery of Uranus

". . . very different from any comet I ever read any description of or saw."
Nevil Maskelyne, Astronomer Royal, April 1781

On March 13, 1781, amateur astronomer William Herschel discovered a new planet. It was completely unexpected.

For as long as anyone could remember, for the thousands of years of recorded history, the solar system ended at Saturn. Mercury, Venus, Mars, Jupiter, Saturn, the Sun, and the Moon slowly shifted their positions against the background stars. That was why the Greeks had named these objects *planētes asteres*—"wandering stars." The telescope, invented early in the seventeenth century, showed some previously invisible features and moons of the known planets. But still no one suspected that planets lay beyond Saturn.

That's why it came as such a surprise to Herschel when he saw a small disk in his telescope on the evening of Tuesday, March 13, 1781. The quality of his eyes and his instrument told him that this was not one of the "fixed stars." When he checked this location between the horns of Taurus and one foot of Gemini on his charts, they did not show a star. Four nights later, when the weather again made observations possible, Herschel saw that his strange object had moved slightly among the stars. Celestial bodies that behaved like this were discovered from time to time. They were called comets. So Herschel published an announcement of this comet, although he noted it was unusual in that it had no tail and showed a distinct rather than a fuzzy disk.

So the actual discovery of this new planet was an accident. That it happened to Herschel was no accident at all.

William Herschel at the age of 46 (1784), three years after his discovery of Uranus (from a crayon copy of an oil painting by L. T. Abbott in the National Portrait Gallery, London)
Courtesy of Special Collections, San Diego State University Library

1

Resourcefulness

William Herschel was born on November 15, 1738, as Friederich Wilhelm Herschel in Hanover, Germany, the third surviving child of an oboist in the Hanoverian Guards military band. Isaac, the father, saw to it that all four of his sons and, when his wife, Anna, wasn't looking, both his daughters not only received excellent musical training but partook of his interest in scientific and cultural matters as well.

William was recognized early as a fine musician. He was admitted as an oboist and violinist to his father's band at age 14.

In 1756, at the onset of the Seven Years' War, the Hanoverian Guards, including the band—father Isaac, older brother Jacob, and William among them—were posted to England for about nine months. Prussia and Austria were the principal adversaries in this war. France allied itself with Austria, so Great Britain, France's worldwide rival in the quest for colonies, sided with Prussia.

As a French attack on Hanover became more likely, the Hanoverian Guards were called home. The band was needed for combat duty to provide the patriotic inspiration and the disciplined beat that encouraged soldiers into battle. William was present on July 26, 1757, during the disastrous Battle of Hastenbeck. The Hanoverian forces were routed. William escaped from the battlefield and rejoined his shattered unit, but the danger, confusion, and the long forced marches in the days ahead convinced him that he was not cut out for army life. His father agreed and arranged for his discharge from the band.[1] William and his older brother Jacob escaped just in time. The retreating Hanoverian Guards were finally trapped and forced to surrender in September. For the next two years, while they occupied Hanover, the French held the Guards captive—band and all—in a military encampment. Isaac was essentially lost to his family for that period and never recovered his health.

Late in 1757, William, at the age of 19, and his brother Jacob left Germany for England. There William copied music while Jacob gave lessons and performed as opportunities permitted. Two years later, upon the defeat of the French, Jacob returned to a musical career in Germany. William stayed on in England. He gave music lessons to the Durham Militia band for two years and then became an itinerant musician, performing in concerts, teaching, and composing in the style of Haydn. He anglicized his name to William. By 1762 he was doing so well that he could send money home to help support his family. But this kind of musical career was too precarious. He sought more secure employment.

Organs were just being installed in English churches, and organist-choir directors were needed. There was one obstacle: William had never played the organ. He began to practice wherever possible, and then he applied for the post of church organist in Halifax. The competition was stiff, including one applicant with extremely agile fingering. William decided on a stratagem. English organs did not yet have pedals for bass notes, as continental organs did. So, unseen, William placed lead weights on the organ keyboard, thus achieving a fullness of sound that impressed and mystified the judges.[2] William won the position, but he stayed in Halifax only three months. Late in 1766 he was appointed organist and choir director for the Octagon Chapel in the resort city of Bath. Music students flocked to him because of his skill and amiability, and he found himself giving as many as 46 lessons a week. With two full-time jobs, he was now earning a good living.

In 1767 William's father died, and his musically talented younger sister Caroline found herself trapped as cook, seamstress, and cleaning woman for her family. William tried to extricate her. In 1772 he finally persuaded his mother to allow his sister to return with him to England by providing enough money so that his mother could hire a servant to replace Caroline. Caroline was forever grateful.

She went from managing one household to managing another that was even more demanding. But now she was happy. She was studying music. Then, in 1773, William plunged into the hobby of astronomy. It began as an effort to improve his musicianship. To help himself understand harmonics, he studied mathematics. After spending 14 to 16 hours a day with music, he would relax at night by solving calculus problems. Math got him interested in optics. Optics got him interested in astronomy. He bought some lenses and assembled a telescope to view the heavens but was very disappointed. He wanted to see more and realized that a larger instrument would be required. But in his day, refracting telescopes were small because the lenses caused colored haloes to appear around every star.[3] The bigger and thicker the objective lens, the worse the problem was. The less the lens curvature, the less the problem of chromatic aberration but the longer the telescope became. Herschel fashioned one 30 feet in length and abandoned it as unwieldy.

So he turned to reflecting telescopes. The large light-gathering lens for a refracting telescope has to be ground on both sides and needs a perfect interior as well, whereas the mirror for a reflecting telescope requires the polishing of only one surface. In fact, the mirror for a reflecting telescope need not be made of glass. Perhaps, thought Herschel, he himself could fashion a decent-size mirror with acceptable reflective qualities in order to collect more light. He experimented with metal alloys until he found one good enough to serve as the mirror of a reflector. His younger brother Alexander, a cellist, had joined him at Bath in 1766, and brother Jacob had returned for a visit to enjoy the musical opportunities that William could arrange. Alexander assisted with the new project, contributing his valuable engineering skills. Caroline also joined in faithfully, partaking in every aspect of the work, despite the shambles it made of their house. During the day, Herschel's home was a school of music. Suddenly, it was a telescope factory as well. During the night, it was an observatory.

THE HERSCHEL SCHOOL OF MUSIC AND TELESCOPE FACTORY

To make a telescope satisfactory to him, Herschel needed to craft a better instrument than had yet been made. He wanted to make a large telescope, so he rejected the Galilean type of refractor, which used a lens to gather light, because he judged that a large lens could not in his day be ground well enough. To gather more light, he turned to reflecting telescopes of the kind that Newton had pioneered a century earlier. He experimented with different metal alloys in an attempt to find one of high reflectivity, fair resistance to tarnishing, and reasonable ability to hold its shape through daily temperature changes.

He turned his entire house into a telescope laboratory, much to the dismay of his sister Caroline: "... it was to my sorrow that I saw almost every room turned into a workshop." Telescope tubes and stands were fabricated in the drawing room. An optical lathe for grinding eyepieces took over a bedroom. The kitchen became the smelter.

On one occasion, in August 1781, a hot mirror mold (made of horse dung) broke, spilling molten metal that cracked the flagstone floor and sent pieces flying like shrapnel in every direction. The Herschels and their workmen fled for their lives. Herschel fell exhausted on a heap of bricks. Miraculously, everyone escaped injury.

In spite of occasional mishaps, Herschel found an acceptable speculum metal from which to make his mirrors—71 percent copper and 29 percent tin. In 1778 he completed an excellent 6.2-inch telescope with a focal length of 7 feet (frequently referred to as the 7-foot telescope) and used it to make his second and third complete surveys of the heavens.

In 1782, a year after Herschel found Uranus with

Herschel's house at 19 New King Street, Bath. It was here on March 13, 1781, that Herschel discovered Uranus.
William Herschel Society, Bath, England

this telescope, it was taken to Greenwich for comparison with the telescopes at the Royal Observatory. Nevil Maskelyne announced that Herschel's telescope was clearly superior.

Astronomy became William's passion. He cut back on his music teaching. English weather being what it is, every clear night had to be used to its fullest. No time was to be lost. William would not pause from telescope making or observing for meals. So, while he worked, Caroline placed morsels of food in her brother's mouth. William also didn't seem to need much rest. Caroline wrote: "If it had not been sometimes for the intervention of a cloudy or moon-light night, I know not when my Brother (or I either) should have got any sleep . . ."[4]

One evening in December 1779, Herschel was in front of his house observing the Moon with his 6.2-inch (7-foot focal-length) reflector when a man came up and asked if he might have a look. Herschel as always was very accommodating. The man later introduced himself as Dr. William Watson. He and his father, physician to the king, were both members of the Royal Society. A deep lifelong friendship began.

Perhaps a Comet

By March 13, 1781, William Herschel, with help from the Watsons, was beginning to become known in English scientific circles. He had completed a survey of the skies with a 4.5-inch reflector in which he cataloged every star of 4th magnitude or brighter. (All these stars were visible to the naked eye, but they were up to 100 times fainter than Sirius, the brightest nighttime star.) That "review," as Herschel called it—the first of four he was to complete in his lifetime—and reports about the size and quality of his telescope had generated much interest. If anyone was most likely to be the first person since the beginning of recorded history to discover a planet, it was William Herschel.

And he did. It happened during his second great all-sky survey. He was using the 6.2-inch reflector he had built after his initial test search. "In the quartile near Zeta Tauri," he wrote in his observing journal, "the lowest of two is a curious either nebulous star or perhaps a comet. A small star follows the comet at ⅔ of the field's distance."[5] For more than a month Herschel continued to think the object he found was a comet. Previous comet orbits as computed by Newton, Halley, and others were long ellipses resembling parabolas. But a parabola failed to correctly predict where Herschel's newly found object would be located a few days or weeks later. Its orbit was not cometlike.

Nevil Maskelyne, the English astronomer royal, was the first person on record to suspect that the object was a planet. He received news of the discovery from Dr. William Watson, Herschel's friend in Bath. After observing for three nights, he wrote to Watson on April 4, 1781: "[The object's motion] convinces me it is a comet or a new planet, but very different from any comet I ever read any description of or saw."

On April 23, he wrote to Herschel:

Nevil Maskelyne, astronomer royal and friend of William and Caroline Herschel
Royal Greenwich Observatory

I am to acknowledge my obligation to you for the communication of your discovery of the present Comet, or planet, I don't know which to call it. It is as likely to be a regular planet moving in an orbit nearly circular round the sun as a Comet moving in a very eccentric ellipsis. I have not yet seen any coma or tail to it.[6]

The mathematical proof came independently that summer from Anders Johan Lexell, a Swedish astronomer working at Saint Petersburg in Russia, and from Jean Baptiste Gaspard Bochart de Saron and Pierre Simon de Laplace in France. When they fitted the observations to a circular, more planetlike orbit, the object moved more nearly as predicted. Laplace and Pierre François André Méchain calculated the first elliptical orbit for Uranus in 1783.

SONG OF THE HEAVENS: CAROLINE HERSCHEL (1750–1848)

"I never forgot," wrote Caroline Herschel, "the caution my dear Father gave me against all thought of marrying, saying as I was neither handsome nor rich, it was not likely that anyone would make me an offer, till perhaps, when advanced in life, some old man might take me for my good qualities."[1]

So Caroline dreamed of a career to keep her from being dependent on family members. She had a good singing voice. She came from a musical family. But her mother disapproved of careers for women. Anna even frowned on the liberal education that Isaac, the father, gave their sons. It would inflate their expectations. She did her best to limit her two daughters to the purely practical matters of cooking, sewing, and household management. Isaac quietly slipped in a few language and music lessons for the girls whenever he could.

But when Isaac died in 1767, Anna Herschel relegated her younger daughter Caroline, age 17, to the position of household drudge. Five years passed slowly and unhappily. Finally, her brother came to rescue her and brought her to England to live. He bought her freedom by providing money

Caroline Herschel at the age of 79 (1829)
National Maritime Museum Greenwich

for their mother to have a paid servant. Wilhelm— he called himself William now—had managed to establish himself in the resort city of Bath as a church organist, music teacher, and concert performer. Caroline was forever grateful to her brother and plunged into intensive training under his direction.

Caroline's day began with breakfast at 7 A.M. or earlier, much too early to suit her. On the breakfast

[1]This and a second quotation from Caroline are from Constance A. Lubbock, ed., *The Herschel Chronicle* (Cambridge: Cambridge University Press, 1933), pp. 45, 137. The spelling and punctuation of this quotation have been modernized. Caroline had mild scarring from smallpox.

Half a year after he found it, Herschel's comet was acknowledged by most astronomers to be a large new planet. Its calculated orbit placed it more than twice as far from the Sun as Saturn. Herschel had doubled the size of the solar system.[7] The disk of the new planet, as measured by Herschel, was a little over 34,000 miles (54,700 kilometers) across, more than four times the diameter of the Earth. Herschel's estimate was not too far from the modern measure of 31,800 miles (51,200 kilometers).

English astronomy, which had generally led the way since Newton a century earlier, now had a brilliant new discovery in its crown and an amateur scientist who was becoming the greatest astronomer of his age.

The Royal Society, England's venerable scientific organization, gave Herschel its coveted Copley Medal in 1781 and made him a member.

---

table, from her brother, were "Little Lessons for Lina"—mathematics problems for her to solve and discuss with William over breakfast. Following breakfast, Caroline took cooking lessons from the housekeeper, whose job she would soon have. Then followed a lesson in English and then another in math, emphasizing bookkeeping. Next she practiced music, playing the harpsichord while singing with a gag in her mouth to improve her projection and enunciation. This was the first of two or three music lessons each day. Finally, she was allowed some relaxation, by which William meant the study of astronomy. That was Caroline's daily schedule up till noon.

Within five years, Caroline became a noted soloist. But her career didn't last long. William was seized by a dangerous and demanding hobby— astronomy. He started building telescopes. Larger and larger telescopes. And observing with them. One night he fell off the scaffolding and narrowly escaped grave injury. Thenceforth Caroline stayed up nights with him, at first taking down observing notes, then making her own observations. During the day, after a full night of observing, she ran the household while her brother taught music students and performed. She also entertained guests, helped her brother prepare astronomy papers for publication, and polished telescope mirrors. Her singing career was over.

While helping William to prepare a telescope for

observations after a snowfall on New Year's Eve 1783, she slipped on the ice and fell against a large metal hook used for lifting the telescope. The hook pierced her right leg six inches above the knee. Her brother and a workman extricated her, but "not without leaving near 2 oz. of my flesh behind." The doctor told her that her injury would have kept a soldier in the hospital six weeks, but she was back in action within days, rejoicing that her incapacity had not cost her brother any good observing weather.

But the turning point had already come. In 1781 her brother had discovered the planet Uranus. William Herschel became world-famous. The next year, at age 44, he abandoned the profession of music and turned all his energies to astronomy.

Caroline Herschel followed her older brother into a new career. The king allotted her a subsidy of 50 pounds a year, making her the first woman to be a professional astronomer. She never married, but the distinguished French astronomer Joseph Lalande made two visits to see Herschel's giant telescope at Slough, "possibly as much because of Caroline as the 40-foot."[2] In eleven years, she discovered eight comets and what we now know is a satellite galaxy of the Great Spiral in Andromeda.

[2]Colin Roman, *Astronomers Royal* (Garden City, New York: Doubleday, 1969), p. 90.

Royal Observatory Greenwich about 1781, when Herschel discovered Uranus and his telescope was brought to the astronomer royal for inspection
National Maritime Museum Greenwich

On its own initiative, the society waived his annual dues so that he would have extra money to continue his research.

Herschel's achievement in discovering Uranus by recognizing its tiny disk as different from the stars stunned the astronomical community. Admirers urged King George III to give Herschel an annual subsidy so that he could devote all his time to science. But others could not believe the claims Herschel made for his instruments. They urged that the 6.2-inch telescope Herschel had made for his second sky survey—the telescope he was using when he found Uranus—be inspected to see if he could possibly be telling the truth about its performance and the discoveries he was reporting, which they could not verify with their own instruments.

In late May 1782, Herschel visited the king and showed him the heavens through his telescope. By mutual agreement, Herschel then stopped by the Royal Observatory at Greenwich so that the astronomer royal and other astronomers could have a look. The word was quick in coming. Maskelyne and other English astronomers pronounced the telescope far superior to those owned by the Royal Observatory or any others they had tested.

Herschel immediately wrote home to his sister to share the news:

> These two last nights I have been star-gazing at Greenwich with Dr. Maskelyne and Mr. Aubert [Alexander Aubert, a prominent amateur]. We

have compared our telescopes together, and mine was found very superior to any of the Royal Observatory. Double stars which they could not see with their instruments I had the pleasure to show them very plainly, and my mechanism is so much approved of that Dr. Maskelyne has already ordered a model to be taken from mine and a stand to be made by it to his reflector. He is, however, now so much out of love with his instrument that he begins to doubt whether it *deserves* a new stand.[8]

A little later Herschel visited Aubert's observatory for a similar side-by-side comparison. It was no contest. "I can now say," he wrote to his brother Alexander, "that I absolutely have the best telescopes that were ever made."[9] Self-taught astronomer and optician William Herschel was not only making the best telescopes in the world, he was also using them in a rigorous observing program like none ever attempted.

Herschel pretty much ignored the fame that followed his discovery, but he was bothered by the impression that the finding was an accident. Because his second all-sky survey, begun in August 1779, had been both comprehensive and precise, Herschel felt any object he found was not due to luck. He was recording the position and brightness of every star down to magnitude 8, about 40 times fainter than the dimmest star in his first all-sky survey. The faintest of these stars was five times too dim for the best human eye to see and ten times fainter than Uranus. As he later explained:

> It has generally been supposed that it was a lucky accident that brought this new star to my view; this is an evident mistake. In the regular manner I examined every star of the heavens, not only of that magnitude but many far inferior, it was that night *its turn* to be discovered. I had gradually perused the great Volume of the Author of Nature and was now come to the page which contained the seventh Planet. Had business prevented me that evening, I must have found it the next, and the goodness of my telescope was such that I perceived its visible planetary disc as soon as I looked at it . . .[10]

The Need for a Name

By the fall of 1781, six months after the discovery, most astronomers had convinced themselves that Herschel's object was a planet, and the need for a name seemed urgent. How should a new planet be named? There was no precedent. No planet had ever been discovered in the course of recorded history.

Sir Joseph Banks, president of the Royal Society, urged Herschel to make a suggestion, lest the French, who had helped to prove that the new object moved like a planet rather than a typical comet, seize the glory:

> Some of our astronomers here incline to the opinion that it is a planet and not a comet; if you are of that opinion it should forthwith be provided with a name [or] our nimble neighbours, the French, will certainly save us the trouble of Baptizing it.[11]

Herschel ignored the request. Meanwhile, the French were indeed engaged in bestowing a name. Joseph Jérôme Le Français de Lalande, one of the orbit calculators, generously and unpolitically suggested that the new planet be named Herschel.

Another proposal came from Johann Elert Bode, editor of the *Berliner Astronomisches Jahrbuch* (Astronomical Yearbook of Berlin). He suggested the name Uranus, because Uranus was the father of Saturn, just as Saturn was the father of Jupiter and Jupiter the father of Mars, Venus, Mercury, and Apollo (the Sun). Uranus was also god of the sky and husband of Earth. The name Uranus would thus keep all the planets part of one mythological family.

These two proposals and many others—including Neptune, Cybele (wife of Saturn), Astraea, and Minerva—were published and debated, with still no response from Herschel. The Royal Society, whose name told of its patronage, saw a good public relations opportunity slipping away. Intent on securing royal assistance for Herschel's research in any case, the society may have encouraged King George III to believe that Herschel wanted to name the new planet after him, while telling Herschel that the king wanted to give him an annual subsidy and that it would therefore be appropriate for him to name the planet after George.

These machinations were delayed somewhat by the insistence of some skeptical society members that Herschel's telescope first be tested. Once the verdict was in, Herschel wrote the king to ask if he might honor him in the planet's name. The royal pension for Herschel came through in the summer of 1782. His subsidy of 200 pounds a year was two-thirds the salary of the astronomer royal, and Herschel had only one official duty: to live close enough to Windsor Castle so that he could show the king and his family the heavens from time to time.[12] This stipend was livable but not comfortable. Herschel was not overwhelmed by the king's munificence. He was earning about twice that amount in Bath by teaching music and making telescopes. Nevertheless, he accepted the appointment because he welcomed the opportunity to devote all his energies to astronomy and because he could continue to supplement his income through the sale of telescopes. The king, in fact, immediately ordered five 10-foot focal-length telescopes to dispense as gifts.

Herschel soon wrote an effusive letter of praise to King George, proposing that the new planet be called Georgium Sidus (George's Star) because of the king's generosity to him and to science and because the name would remind people that the planet was found in England during the reign of George III.

The name was instantly unpopular wherever George III did not reign. Astronomers resented the name's political symbolism and cumbersome length, and they pointed out that George's Star was not a star. Each coun-

William Herschel holds a diagram of the Georgian Planet (Uranus), orbited by the two moons (Oberon and Titania) that he also found.
Courtesy of Special Collections, San Diego State University Library

try went its own way. The new planet was Uranus in Germany; it was called Herschel (and then gradually Uranus) in France; and for the next 60 years in Britain it was known as the Georgian Planet. Bode's recommendation finally prevailed, but the new planet remained officially "The Georgian" in Britain until after the discovery of Neptune and through the 1847 publication of the *Nautical Almanac* for 1851.

How serious was Herschel about the planet name he had suggested? His gracious letter to the king sounds sincere. Yet when Bode wrote in 1783 to congratulate him and to argue that Uranus was a better name, Herschel responded: "When I named it Geo[rgium] Sidus I hardly expected that this name would become generally accepted, because we already know by experience that the first names of the satellites of Jupiter and Saturn were soon changed."[13] Was Herschel just being diplomatic and standing aloof from partisan disputes, or was he acknowledging that his proposed name was a gesture to his benefactor and of little interest to him?[14] There is no way to tell. Whatever Herschel's intentions, he left no evidence that he wanted the planet named after himself.

The Amateur Turns Professional

In accordance with the terms of his royal pension, Herschel immediately moved close to Windsor. In 1786 he settled in Slough, within sight of the castle, where he spent the rest of his life. He gave up music and devoted his remaining 36 years to astronomy.

Even in his occasional duty as astronomy tutor to the royal family, Herschel demonstrated his characteristic creativity and resourcefulness. For example, if the court was anxious to observe but the weather was uncooperative, Herschel would rig up a model Saturn at a distance and illuminate it by a lantern so that the expectant royalty could look through the telescope anyway and enjoy at least a simulated view.[15]

Herschel was 42 years old when he found Uranus, an age when most scientists' major discoveries are behind them. But, for Herschel, 40 years of sustained scientific productivity lay ahead. From 1781 until 1821, when his health finally failed, he led the way in virtually every branch of astronomy.

In 1788, at age 49, Herschel married a 37-year-old widow named Mary Pitt, and in 1792 their only child, John, was born. Even as a boy, John began helping his father and Aunt Caroline, and he grew up to be one of the most renowned astronomers of his age. In the meantime Mary saw to it that William now took time for meals and vacations. For a time, Mary's place in William's affections strained her relations with Caroline, who had so selflessly devoted her life to her brother. But eventually both women, to their credit, came to like each other.

A serious illness in 1808 caused Herschel, at age 70, to cut back his observing schedule somewhat, but he remained active until 13 years later, when his health began to decline steeply. In June 1821, Caroline served as his observing assistant for the last time. Herschel died on August 25, 1822, just short of 84 years old. His life was almost exactly the length of one revolution of the planet he had discovered. Uranus had nearly returned to the place in the skies where it had stood at his birth.

Throughout most of the eighteenth century, research in astronomy meant the confirmation and application of Newton's work. Professional astronomers often became bureaucrats, endlessly logging the precise time that the Sun, Moon, stars, and planets crossed the north-south line in their transit telescopes so that they could improve the measurement of time, determine geographical longitude, and refine the orbits of the planets.

It was amateur astronomer William Herschel and three generations of others like him who changed the course of astronomy and revitalized

HERSCHEL'S TELESCOPES

The discovery of Uranus brought Herschel a flood of inquiries about his instruments and techniques. Only then did it begin to dawn on him that he, William Herschel, an amateur isolated in Bath, England, was the best-equipped astronomer in the world. Despite Herschel's sale of telescopes, the world did not catch up with him in his lifetime. Because other astronomers did not check his observations on stars and faint nebulae, they could not appreciate his work on the structure of the heavens.

After he came under patronage from the king, Herschel began work on the largest telescope the world had ever known—a 40-foot-long instrument with a 48-inch mirror. This endeavor was subsidized by an initial grant of 2,000 pounds from King George III. A second 2,000 pounds followed to complete the job, and 200 pounds a year was allocated to maintain the instrument. Herschel supervised every detail of the work, which sometimes required 40 workers. It took 12 men working together to handle the grinding and polishing tool. Herschel had three mirrors cast, each weighing almost a ton, before he was reasonably satisfied. The telescope was three and a half years in the making.

The 50-foot-high scaffolding for the huge tele-

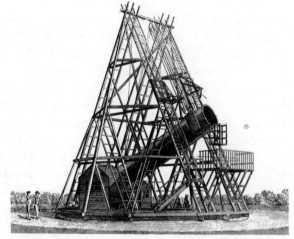

Herschel's 40-foot (48-inch) reflecting telescope
Royal Greenwich Observatory

scope attracted many visitors, and Herschel lost much time showing distinguished guests, including the king, his work. People especially liked to walk through the 40-foot-long tube as it lay on the ground. One day the king brought the archbishop of Canterbury to see the project and led him it. For 60 years after Herschel discovered Uranus in 1781, progress in astronomy came almost entirely through the efforts of amateurs.

Herschel was acclaimed around the world and is still recognized as the father of stellar astronomy. He made four highly successful complete surveys of all the stars visible through his ever-improving telescopes. He discovered the existence of binary star systems—stars revolving around one another. He realized that stars were much farther away than previously thought. He pioneered the study of the shape and extent of the stellar universe, thus founding the field of observational cosmology. He proved that the Sun was moving in space, toward a point in the constellation Hercules. He discovered infrared light. All these breakthroughs came after he found the planet Uranus. "The history of astronomy," wrote historian Michael Hoskin, "knows great telescope builders, great observers, and great theorists; but only William Herschel falls indisputably into all three categories."[16]

---

through the tube, saying, "Come, my Lord Bishop, I will show you the way to Heaven."[1]

Completed in 1789, the 48-inch reflector remained the largest telescope in the world for more than half a century until William Parsons, the lord of Rosse, made a 72-inch reflector in 1845. Herschel's 48-inch telescope was larger than any optical telescope in operation in Britain today.[2]

On his second night of viewing through his 48-inch telescope, Herschel discovered a sixth moon of Saturn (Enceladus). Three weeks later he discovered a seventh (Mimas). Yet Herschel was disappointed with this instrument. It was clumsy to operate, the English weather was seldom good enough to allow the large mirror to perform well, the mirror distorted under its own weight when tilted, and it tarnished rapidly. With only a hundred hours a year of truly excellent seeing conditions,[3] Herschel had at last exceeded the limits of contemporary technology and the weather of his adopted country. He returned to his 20-foot focal-length (18.8-inch) reflector (the "large 20-foot," completed in 1783), one of the finest telescopes ever made. He used it constantly for 20 years.[4] The 40-foot monster was left mostly for visitors to gawk at.

Another of Herschel's finest telescopes was the 25-foot-long, 24-inch-diameter reflector ordered by the king of Spain for the Madrid Observatory. Herschel was paid 3,150 pounds. Unfortunately, most of the people who bought a Herschel telescope did so for prestige and as a work of art. Little astronomical use was ever made of this exquisite instrument, and it was destroyed in the Napoleonic wars.

The giant 40-foot telescope stood erect until 1839, when the rotting timbers of the scaffolding compelled John Herschel to have the instrument taken down. The tube, mirror in place, was laid on the grass, and the family gathered inside the tube for a requiem, using music composed by John. The tube was then sealed against the elements. A tree felled by a storm in the 1860s crushed most of the tube, but the mirror and a fragment of the tube are still displayed at Herschel's home in Slough today.

---

[1]Mary Cornwallis Herschel, *Memoir and Correspondence of Caroline Herschel* (New York: D. Appleton, 1876), p. 309.

[2]Pointed out by Michael Hoskin, *William Herschel and the Construction of the Heavens* (New York: W. W. Norton, 1963), p. 26.

[3]According to Herschel, cited by Henry C. King, *The History of the Telescope* (Cambridge, Massachusetts: Sky Publishing, 1955), p. 133.

[4]His son John Herschel refurbished this telescope under his aged father's supervision, resurveyed his father's nebulae and star clusters with it, and then took this instrument to the Cape of Good Hope to survey the southern skies, his most famous contribution to astronomy.

CHAPTER 2

The Fervor for New Planets

"Can one believe that the Founder of the Universe left this space empty? Certainly not."
Johann Elert Bode
(1772)

Before the discovery of Uranus, no one thought that there were any planets beyond Saturn. Almost no one thought that there were unknown planets closer to the Sun than Saturn. But with the discovery of Uranus in 1781, horizons suddenly broadened. Maybe there were other planetary bodies in the solar system yet to be found. Attention focused on something called Bode's Law.

Popularized by the same man who provided Uranus with its name, Bode's Law began as a curiosity—the relative distances of the planets from the Sun expressed as an arithmetic progression. Two-thirds of a century later, it was once again just a curiosity. But during the intervening 65 years, Bode's Law was a major driving force in solar system research. The distance of newly discovered Uranus was in good agreement with the law's numerical sequence, and this gave the progression credibility.

Titius' Approximation

Bode's Law, however, is neither Bode's nor a law. More properly, it should be (but seldom is) known as the Titius-Bode Rule. In 1766, Johann Daniel Titius, a mathematics professor at Wittenberg, translated from French to German a work by Swiss naturalist Charles Bonnet entitled *Reflections on Nature.*[1] Bonnet was trying to show the handiwork of God in the order of nature, and he began with the example of the solar system.

14

The example was too diffuse, however, and Titius decided to help Bonnet by quietly interposing a paragraph that offered an intriguing progression in the distances of the planets from the Sun.[2] His finding can be stated like this:

- Start with the series 0, 3, 6, 12, 24, 48, 96, . . . in which, beginning with 3, each succeeding number is twice the one before it.
- Add 4 to each number.
- Divide each number by 10.[3]

Johann Daniel Titius
Archive of the Archenhold Observatory

The result is a good approximation of the distances of the then-known planets from the Sun in astronomical units (an astronomical unit is the mean distance of the Earth from the Sun—now set at 92.96 million miles, or 149.6 million kilometers).

DISTANCE FROM THE SUN
(in astronomical units)

	Titius-Bode Law		*Actual as of 1790*
Mercury	$\dfrac{0+4}{10}$	= 0.4	0.39
Venus	$\dfrac{3+4}{10}$	= 0.7	0.72
Earth	$\dfrac{6+4}{10}$	= 1.0	1.00
Mars	$\dfrac{12+4}{10}$	= 1.6	1.52
———	$\dfrac{24+4}{10}$	= 2.8	(nothing known)
Jupiter	$\dfrac{48+4}{10}$	= 5.2	5.20
Saturn	$\dfrac{96+4}{10}$	= 10.0	9.54

Titius noted that his numbers worked well except for a gap between Mars and Jupiter where there was no known planet. "But did the Lord Architect leave that space empty?" he asked. "Not at all," he answered.[4] This space must be occupied by undetected satellites of Mars and perhaps of Jupiter.

This exercise by Titius, buried in another author's book, would probably have attracted no attention had not Johann Elert Bode happened across it. Bode was a young, energetic, self-taught astronomer who, at the age of 21, had published a very popular introduction to the heavens.[5] In 1772 he had just been hired by the Berlin Academy of

Johann Elert Bode
*Archive of the Archenhold
Observatory*

Sciences to work on its annual astronomical almanac. The academy relied on this publication for revenue, but it was selling poorly. Bode quickly transformed it from a money loser to a high-profit item by correcting the publication's inaccuracies and by supplementing the usual numbers, dates, and times of an astronomical almanac with astronomy and general science news from around the world. He thereby made the *Berliner Astronomisches Jahrbuch* (Astronomical Yearbook of Berlin) into a major reference work. He continued as editor from 1774 (age 27) until his death in 1826.

Bode came across the planetary distance relationship in the first German edition of Bonnet, translated by Titius in 1766. Titius inserted it into the text of the book without a note, so that it appeared to be by Bonnet.[6] Bode reprinted the relationship with only a few very minor changes in phrasing and without credit to Bonnet or Titius in the second edition (1772) of his introductory astronomy book. "Can one believe," said Bode, "that the Founder of the Universe left this space empty? Certainly not."[7] Instead of satellites of Mars or Jupiter in the gap, however, Bode proposed that a major planet lay undiscovered there and enthusiastically promoted this planet spacing relationship every chance he got.

Bode continued to fail to credit Titius in the next two editions of his popular astronomy book and in his 1778 technical survey of astronomy as well. He finally acknowledged Titius as the source for Bode's Law in 1784, three years after Uranus was discovered, when he wrote a monograph on the new planet. Bode pointed out that if his rule about planet distances had been carried one step further, it would have predicted a planet at 19.6 astronomical units and that Uranus was actually found at 19.2. Suddenly, this curious rule acquired the aura of credibility. And this formulation of planet spacings became tightly coupled to Bode's name.

Celestial Police

Based on the apparent validation of Bode's Law by Uranus, Bode began plans to search for the planet that the law predicted should lie between Mars and Jupiter. The object had to be faint, or as close as it was, it would be plainly visible to the unaided eye.

Bode's confidence that a planet existed in the Mars-Jupiter gap was matched by that of the Hungarian astronomer Franz Xaver von Zach. On September 21, 1800, Zach convened a meeting of astronomer friends in Lilienthal, Germany, at the home of the town's chief magistrate, Johann Hieronymus Schröter.[8] The village of Lilienthal was about 12 miles (20 kilometers) from Herschel's birthplace in Hanover. Schröter was an amateur astronomer who made outstanding telescopes himself and who had purchased a superb reflector with a 27-foot focal length. This gave him the largest telescope on the European continent. Schröter was one of the few people to use a Herschel telescope for research rather than display.

King George III, ruler of Hanover as well as Great Britain, subsidized both Herschel and Schröter. The king encouraged Schröter's research by buying all of Schröter's instruments and leaving them in his hands until his death. He also provided money for Schröter to hire an assistant, just as Herschel received a stipend for Caroline. Schröter is regarded as the father of selenography for his long-term work on the features of the Moon.

Franz Xaver von Zach
Archive of the Archenhold Observatory

The six astronomers[9] who met at Lilienthal adopted a plan previously proposed by Lalande to find the missing planet. Since all the known planets lay near the ecliptic, within the zodiacal band, they reasoned that the undiscovered planet might also be found there. So they divided the sky into 24 sections along the zodiac and assigned the search of each 15-degree section to an astronomer they felt would want to participate in this project. Zach dubbed the group the Lilienthal Detectives. Letters went out to astronomers all over Europe notifying them of their assignments.

It was too late. Before the letter to Giuseppe Piazzi arrived at the observatory he directed in Palermo, Sicily, he had made and reported a puzzling discovery.

Piazzi was a priest and mathematician by training. When he was appointed a professor at Palermo in 1780, he was encouraged to establish an observatory. In the late 1780s he went to England so that he could buy the best instruments available. He spent time with the astronomer royal and then went on to visit William Herschel. While observing with Herschel, he fell off the scaffold and broke his arm. Astronomy can be a dangerous profession.

Johann Hieronymus Schröter
Archive of the Archenhold Observatory

Back in Sicily with his new equipment, he launched a lengthy, painstaking, but valuable project to determine the exact astronomical coordinates of several thousand stars. He was 11 years into this project when, on the night of January 1, 1801, the first day of the nineteenth century, he noticed a very faint, starlike object in Taurus that wasn't in the catalog he was checking.

He observed it again the next night and the night after that. The object was moving slowly against the background stars. A few more days of observations assured him that the object was indeed a member of the solar system—but what was it? He began to notify colleagues. To Lalande and Bode he described his discovery as a comet. But in his letter of January 24, 1801, to his close friend Barnaba Oriani, director of the Brera Observatory in Milan, Piazzi confided:

Giuseppe Piazzi
Courtesy of Special Collections, San Diego State University Library

I have announced this star as a comet, but since it is not accompanied by any nebulosity and, further, since its movement is so slow and rather uniform, it has occurred to me several times that it might be something better than a comet. But I have been careful not to advance this supposition to the public. I will try to calculate its elements when I have made more observations.[10]

On February 11, Piazzi fell seriously ill, and before he had recovered, the object had moved too close to the Sun's position to be observed. Mail delivery could be absurdly slow in those days, and it was two or three months before Oriani, Bode, and Lalande received their letters. Bode immediately suspected that Piazzi's discovery was the planet the Detectives had hoped to find, and he spread the news. Zach was delighted.

For philosopher Georg Wilhelm Friedrich Hegel, however, the news was embarrassing. Early in 1801, before word of the discovery was received, he had published his *Dissertatio Pholosophica de Orbitis Planetarum* (Philosophical Dissertation on Planetary Orbits), in which he proved by logic that only the seven known planets could exist and that Bode's Law was absurd.

But now there was a serious problem with the object that Piazzi had found. By the time Bode, Lalande, and Oriani received notification and Piazzi was well enough to resume work, the object was lost. Piazzi had observed it over a period of 41 days, which covered 3 degrees of its orbit. Such a small percentage of an orbit made the calculation of the object's position after it emerged from the brightness of the Sun extremely uncertain. True, it was less uncertain than the position of Uranus after half a year of observation (by which time it had shown itself to be a planet), but Piazzi's object was much harder to find because it was of 8th magnitude, about ten times fainter than Uranus. It was on the fringe of visibility with most of the telescopes available, and it exhibited no telltale planetary disk. Searches were in progress everywhere, but the new object eluded detection, even by Herschel. Astronomers began to despair.

But then a brilliant 23-year-old German mathematician stepped in: Carl Friedrich Gauss. The problem of recovering Piazzi's object fascinated him. In the fall of 1801 he developed a new method of calculating orbits from a minimum of observations. Late that year he delivered his orbit calculation and position prediction to Zach. On December 31, 1801, the last night of the year of its discovery, Zach found the object only half a degree from where Gauss had calculated it would be.[11]

Gauss's calculation also confirmed what had been suspected: The object had the nearly circular orbit of a planet, not the elongated orbit of a comet. And Gauss confirmed that the object's distance from the Sun was 2.77 astronomical units—almost exactly the 2.8 figure that Bode's Law promised.

Carl Friedrich Gauss, later in life, on the terrace of the Göttingen Observatory
Courtesy of Special Collections, San Diego State University Library

Piazzi proposed to name the object Ceres Ferdinandea—Ceres of Ferdinand. Ceres was the Greco-Roman goddess of agriculture and patron goddess of Sicily. Ferdinand was the ruler of Sicily. But the length of the name and its political connotations quickly resulted in the use of Ceres only.

But soon new trouble arose. Ceres was relatively close, showed no appreciable disk, and was extremely faint. It had to be small. Was it too small to be the expected planet?

Too Small to Be a Planet

Then, on March 28, 1802, Wilhelm Olbers found a new object with very similar behavior. Olbers, a physician and skillful amateur astronomer, was also one of the Lilienthal Detectives. He had been searching to recover Ceres and thus had made himself very familiar with the part of the sky (Virgo) where Ceres might be expected to reappear. In fact, Olbers had independently recovered Ceres on January 1, 1802, the night after Zach's success.

Olbers was looking for comets when he saw what he named Pallas (another name for Athena, sometimes regarded as goddess of medicine). His familiarity with that part of the sky allowed him to recognize this faint interloper from among hundreds of stars in his field of view.

The discovery of Pallas complicated matters. Bode's Law predicted one planet at 2.8 astronomical units. Now there were two. And neither was of normal planetary dimensions. Herschel estimated the diameter of Ceres

Wilhelm Olbers
*Archive of the Archenhold
Observatory*

at 162 miles and Pallas at 110. Modern measurements are kinder: 617 and 335 miles (993 and 540 kilometers), but these two bodies are still much smaller than the smallest planet, Pluto, which is about 1,457 miles (2,345 kilometers) in diameter.

Herschel wrote to Piazzi and Olbers to suggest that it was possible to save Bode's Law only by reckoning that Ceres and Pallas were a new class of solar system members, too small to be designated planets. He proposed to call them *asteroids* because of their starlike appearance, which did not readily reveal a planetary disk. The name stuck, even though asteroids have nothing to do with stars. Planetary scientists now prefer to call them *minor planets* or even *planetoids,* the name proposed by Piazzi in 1802 to avoid the "starlike" connotation of *asteroid.*

Olbers accepted Herschel's diminutive measurements for the two new objects and adopted the term *asteroid* in his reply to Herschel in June

---

AN EARLY APPLICATION OF AUTOMATION TO ASTRONOMY
by Ruth S. Freitag, Library of Congress

After Piazzi came across the first (and largest) asteroid in 1801, three more were found by Olbers and Harding in the ensuing six years; then nearly four decades elapsed without any further discoveries. However, once the search was resumed by a new cast of characters armed with better star maps, the numbers began to accumulate, beginning in 1845. From 1847 on, not a year passed without the identification of yet another asteroid or two, or six, or eleven. By mid-1873, more than 130 were known. After considering carefully the hardships that dedicated asteroid seekers had to endure, the Bavarian satirical weekly, *Fleigende Blätter,* announced in its issue of September 19, 1873, the development of an indispensable new accessory:

An Automatic Asteroid Finder

We have succeeded in building an instrument that discovers asteroids, comets, and other unknown celestial bodies by purely mechanical means. The arrangement is as follows: An instrument resembling a telescope, the lens combination of which is designed like that of the camera obscura, is set up facing the sky. This instrument projects all the heavenly bodies within its field of view onto a long strip of paper, which stretches between two rollers and is carried along past the focal point of the instrument, by means of a clock drive, at the same speed as that of the stars in the field. All the known fixed stars, planets, and asteroids are shown as dark spots on this paper strip, exactly like a celestial map. The rest of the paper is chemically treated in such a way that the least amount of light falling upon a spot that is not blackened will set the paper on fire. If this rolling chart of the heavens is correctly oriented, the light from a star will fall upon the corresponding black spot on the map; but if there is a star (asteroid or comet) anywhere in the sky that is not yet shown on the map, i.e., not yet discovered, a point of light will fall on the chemically prepared part of the paper; the paper will catch fire, sparks will fall on a fuse treated with powder that connects with the touch hole of a small cannon, and a shot will go off. By means of the shot the astronomer quietly sleeping in an adjoining room will awaken, jump up, rush to the instrument, and, in the field of view of a telescope set up parallel with the camera-telescope, he will see the newly discovered star. Klinkerfues, Luther, and all the other famous asteroid discoverers need no longer expose

1802. But Olbers, a clever mathematician, offered an intriguing idea in his letter: that Ceres and Pallas might be fragments of a large planet that had been destroyed by internal explosion or by collision with a comet. He proposed that astronomers search the two regions in the sky where the orbits of Ceres and Pallas came closest to intersecting in the hope of finding other fragments, since an explosion might throw debris in all directions, but the orbits of the fragments would bring them back to the site of the explosion at each revolution. The intersections were in Cetus and Virgo.

On September 2, 1804, Carl Ludwig Harding, an assistant at Schröter's observatory and himself one of the Lilienthal Detectives, discovered Juno in Pisces, very close to Cetus, and on March 29, 1807, Olbers, taking his own advice, found Vesta in Virgo.

It was an embarrassment of riches. All four asteroids were small. Even

Automatic Asteroid Finder

themselves to the night air during frosty winter evenings with their comet-seekers and catch colds and chills from their tiresome work. Unfortunately it has not yet been possible to bring this epoch-making instrument to the international exhibition. It can be expected, in view of the large number of small planets that are to be found between Mars and Jupiter, that every quarter of an hour a shot will sound, announcing to the earth's entire population the discovery of yet another new asteroid.

(Translation by Ruth S. Freitag)

Fleigende Blätter, founded in November 1844, poked fun at German social, cultural, and political life with great success during the nineteenth century. Some of its contributing artists, such as Wilhelm Busch and Carl Spitzweg, achieved independent fame. As the years passed, the journal was overtaken by rapidly changing times to which it could not adapt; it expired, finally, in 1944—the only German periodical of its kind to last for a century.

if they were fragments of a larger planet, all of them put together and all the other pieces that could reasonably be imagined to exist were not enough to form one decent-sized primary planet. Today, the total mass of the asteroid belt (3,700 known asteroids and all those remaining to be identified, plus debris and dust) is estimated to amount to less than 5 percent of the mass of the Moon. It would take almost 2,000 such asteroid belts to equal the mass of the Earth. The zone of minor planets continues to fragment and reduce itself to dust through collisions, but it seems never to have contained a planet of major size.

Nevertheless, most astronomers in the early nineteenth century felt that Bode's Law was extremely powerful. Why it worked remained a mystery, but Uranus was found at the distance the law predicted, and a search of the gap given by the sequence produced the discovery of the asteroids. Some astronomers even used Bode's Law to suggest that a planet lay beyond Uranus.[12]

Confidence in Bode's Law was great enough so that when John Couch Adams and Urbain Jean Joseph Le Verrier attempted to calculate the existence and position of a planet beyond Uranus, both adopted the rule's prediction that a trans-Uranian planet would lie at a distance of 39 astronomical units.

Neptune was found where Adams and Le Verrier, using Bode's Law to postulate its distance, had predicted it would be in the sky, but Neptune proved to be closer than expected—30.06 astronomical units—an error in the law of 23 percent. And Pluto, whose mean distance from the Sun is close to where Neptune should have been, is only half the distance that the Law would predict for a ninth planet.

Ironically, the very success of Bode's Law in helping with the discovery of Neptune led directly to its discrediting. Undermined at the peak of its popularity, Bode's Law returned to the status of a curiosity.[13]

CHAPTER 3

Trouble with Uranus

"Who can say but your new star (Uranus), which exceeds Saturn in its distance from the sun, may exceed him as much in magnificence of attendance? Who knows what new rings, new satellites, or what other nameless and numberless phenomena remain behind, waiting to reward future industry . . . ?"

Sir Joseph Banks, president, Royal Society, on awarding Herschel the Copley Medal (November 1781)

Uranus required a major conceptual leap for the public and the scientific world. A new planet. A solar system twice the size previously thought. The law of gravity extending deeper into the universe.[1]

Then, to further complicate matters, Uranus would not cooperate. It seemed to be violating the law of gravity. In 1781, soon after Uranus was discovered, Lexell, Bochart de Saron, and Pierre Simon de Laplace had calculated circular orbit approximations for Uranus which demonstrated that the object, unlike any known comet, was in a planetary orbit. Laplace and Pierre François André Méchain calculated the first elliptical orbit for Uranus in 1783. That orbit allowed astronomers to predict more accurately where Uranus would be. But still there was considerable error.

New calculations were made, some improvement resulted; and most astronomers presumed that more observations would lead to better ephemerides.[2]

They didn't. In fact, Uranus seemed to be increasingly wayward.

When it is closest to Earth, Uranus has a visual magnitude of about 5.5, barely visible to excellent human eyesight under perfect observing conditions without optical aid.[3] With even a modest amateur telescope, however, Uranus is relatively bright, one of the 2,000 brightest objects among the million or so stars visible with a small instrument. Late in 1781, soon after Herschel's newly discovered object was generally acknowledged to be a planet, Bode raised the question of whether Uranus had been seen previously but mistaken for a star. Such prediscovery observations could be very useful for the refinement of orbital calculations. The position of Uranus had been recorded since its discovery for only a little over six months—too small a span of the 84-year orbit to be confident of its size and period. Additional sightings from earlier years could be helpful in reducing the orbital uncertainties.

Those Who Missed

Bode began a search through historical records and found two occasions when astronomers had recorded a star of the proper brightness that was no longer at its marked position. The motion of Uranus, calculated backward along its orbit, seemed to place the planet near enough to those positions to indicate that earlier astronomers had—not surprisingly—mistaken Uranus for a star. Telescopes prior to Herschel's could not have shown astronomers a planetary disk, which would distinguish it from the stars. Even in Herschel's day, famed observers such as Charles Messier identified Uranus only with great difficulty by noting on consecutive nights which starlike object in the proper field of view had moved against the starry background. Unless they had an instrument like Herschel's, an exacting eye like Herschel's, and an observing program like Herschel's where every single object in the sky, even objects ten times fainter than Uranus, would be recorded—unless they had provided themselves with all these advantages—their chances of discovering Uranus were minimal at best.

Herschel had not been looking for a planet at all. He was conducting his all-sky "review" in search of double stars that might be useful for parallax measurements to determine the distances to stars. He was not specifically looking for motion within the star field from one night to the next, which would indicate a nearby object such as a planet or a comet. He did not recognize Uranus because of its motion but because, to his trained eye and with his matchless instrument, Uranus showed a very small disk and therefore did not quite look like a star.

Bode announced that Tobias Mayer had unknowingly seen Uranus in 1756. So had John Flamsteed, the first astronomer royal, as far back as 1690.[4] With observations now spanning almost a century—more than one orbit of Uranus—several astronomers attempted to calculate an orbit that would allow the accurate prediction of future positions. The calculations worked marginally for a few weeks and then steadily faded into uselessness. Other astronomers joined in the hunt through historical archives for earlier sightings. Pierre Charles Le Monnier examined his own records and found that he had seen Uranus nine times before its discovery without recognizing that it was not a star. He was so embarrassed that he published only four.[5]

New orbital calculations were made that tediously incorporated the gravitational perturbations on Uranus by Jupiter and Saturn. The result still wasn't quite right, but it was an improvement, and astronomers turned their attention to other matters, such as how to avoid the guillotine following the French Revolution. Jean Baptiste Gaspard Bochart de Saron, who had computed one of the first orbits for Uranus, was beheaded. He spent his last night in prison calculating a refined orbit for Halley's Comet.

Troublemaker

Later, 22 years after Le Monnier's death, French astronomer Alexis Bouvard reexamined earlier observational records in an attempt to compute a better orbit for Uranus. He published a list of 17 prediscovery sightings[6] of Uranus and 40 years of postdiscovery observations. These tables (1821) provided position markings along the orbit of Uranus that stretched for 130 years, from 1690 to 1820—more than one and a half revolutions for the planet. With so much coverage, it should have been possible at last to calculate an orbit that truly matched the planet's motion.

But using all the data, Bouvard could find no decent fit. It was frustrating. If all the positions were included, the old observations would correspond, but the new ones wouldn't. If the modern observations alone were used in the calculations, then the old observations wouldn't fit. Bouvard saw no alternative but to exclude the prediscovery sightings from his orbital calculation on the grounds that the observations of Flamsteed, Mayer, James Bradley, and Le Monnier could not be as accurate as modern ones.

Bouvard pointed out that 12 of the 17 prediscovery sightings belonged to Le Monnier, including 4 on consecutive nights in January 1769. Bouvard attributed Le Monnier's misfortune less to lack of luck than to his sloppiness and his failure to compare his notes on star positions from one night to the next. A story circulating at the time said that Le Monnier scribbled one of his unrecognized observations of Uranus not in an observing journal but on a paper bag that had contained hair powder.

The story was not true, but Le Monnier was an easy butt of ridicule. He had an irascible personality and fought frequently with colleagues. Besides, it was in Bouvard's interest to cast aspersions on the non-discoverers of Uranus because he needed to discredit their observations in order to calculate an orbit that seemed to fit the planet's motion.

Le Monnier did not recognize Uranus, but the positions he recorded for this starlike object had an average error in longitude of only 6 arc seconds—ten times smaller than the errors that Bouvard wished to attribute to him.

On the surface, it looked like Le Monnier had indeed been inexcusably inattentive and sloppy. In less than a month, between December 27, 1768, and January 23, 1769, he had seen but not recognized Uranus eight times. Yet during this period of time Uranus was just coming out of retrograde motion and was at a stationary point. Le Monnier did not notice the motion of Uranus through the star field because, at that time, Uranus was virtually not moving.[7]

By excluding prediscovery sightings from his orbital calculations, Bouvard was dismissing the work of some of the guiding lights of positional astronomy. The typical observational uncertainties by all four "ancient" astronomers who had seen Uranus prior to its recognition as a planet were seven to ten times too small to warrant such drastic action. As Le Verrier proved later, the "ancient" data were good. Bouvard wrote unhappily:

> . . . I leave to the future the task of discovering whether the difficulty of reconciling the two systems results from the inaccuracy of the ancient observations, or whether it depends on some extraneous and unknown influence which may have acted on the planet.[8]

Having explained his exclusion of older data, Bouvard then used the 40 years of "modern," more reliable data to establish a new, improved orbit for Uranus. There remained discordances of a fraction of a minute of arc, but additional observations, he felt, could be expected to solve the problem. It seemed promising at first.

But additional observations only exacerbated matters. By 1824, the failure of Uranus to operate on schedule was so aggravating that the distinguished German mathematician Friedrich Wilhelm Bessel expressed his opinion that the "mystery of Uranus" would be solved by the discovery of an outlying planet. He began advocating a search for it.

Throughout this period, Uranus was running fast. It was consistently and increasingly ahead of its predicted position. Then, just as the positional errors seemed intolerably large, the problem began to vanish. Uranus was running less and less ahead of schedule. For about two years, 1829 and 1830, Uranus was almost exactly where calculations forecast. Temporarily, astronomers forgot about the planet's many years of unruly and unexplained behavior.

Then, just as complacence was setting in, Uranus fell off schedule again—but this time it was running slow. Slow, and getting steadily worse.

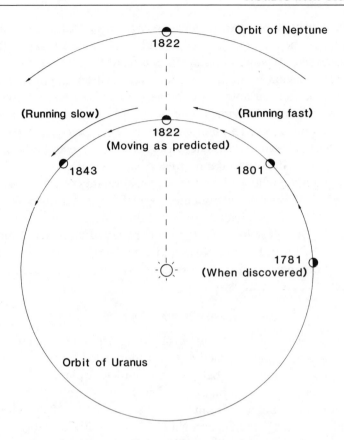

Orbit of Neptune

1822

(Running slow) (Running fast)

1822
(Moving as predicted)

1843 1801

1781
(When discovered)

Orbit of Uranus

From the time of its discovery until about 1821, Uranus moved *faster* along its orbit than provided by calculations using the gravitational attractions of the known planets. Around 1822, Uranus moved as expected. Thereafter, it moved too *slowly.* The reason was an unknown planet beyond Uranus—Neptune.

The effort to explain the problem shifted from observational accuracy to some force acting on Uranus. Several forces were proposed and swatted down. One idea, revived from Descartes, was that a cosmic fluid with vortices filled the space between heavenly bodies. It was discarded, as before, because there was no evidence for such a fluid and no way that it could bring about the perturbations observed.

The second idea proposed that Uranus was being disturbed by a large satellite. This theory was abandoned because, to produce the observed disturbance, the moon would be too big to be undiscovered, and because any moon's period of revolution would be too short to cause such long-period perturbations.

A third proposal was that a comet struck Uranus about the time of its discovery, changing its orbit, so that old and new observations could not be reconciled because the planet now moved in a slightly different path. But this impact theory failed because Uranus was now significantly off schedule according to calculations based solely on modern positional measurements.

The fourth proposal was more tenacious: that the motion of Uranus couldn't be handled adequately by the law of universal gravitation because gravity was not universal. Instead, said some scientists, gravity operated as described by Newton on Earth and from the Sun out to Saturn, but at greater distances the law needed modification. The intractability of Uranus was an example of the breakdown of Newtonian physics. This idea, dating from Newton's time, was proffered whenever a stubborn scientist encountered a phenomenon or developed a theory that didn't immediately seem to reconcile with gravity. Yet in every case over a century and a half, Newton and gravity had prevailed.

Among those who doubted the universal validity of gravity was George Biddell Airy, appointed astronomer royal of England in 1835. For the most part he kept his suspicions quiet, but this bias and his efforts to be the man who fixed Newton's work were to have profound consequences for the lives of several astronomers and the history of astronomy. Yet nearly all of the most productive scientists saw ample evidence that gravity operated in the same way everywhere.

Toward an Explanation

As the 1830s passed, only one proposal to explain the peculiar motion of Uranus remained attractive to those gripped by the problem: that Uranus was being disturbed by a more distant planet.

The idea of unknown planets had surfaced without generating much attention even before Uranus was discovered. In 1758, when Alexis Claude Clairaut, Lalande, and Nicole Reine Étable de Labrière Lepaute had calculated the date when Halley's Comet would pass closest to the Sun on its first predicted return, Clairaut had noted "that a body which travels into regions so remote, and is invisible for such long periods, might be subject to totally unknown forces, such as the action of other comets, or even of some planet too far distant from the Sun ever to be perceived."[9]

Of course, Bouvard had timidly raised the possibility of "some extraneous and unknown influence" on Uranus that was causing his tables for the planet to fail. By 1836 he had embraced this concept rather than blaming discrepancies in the orbit of Uranus on observational errors by earlier astronomers.

Apparently Bouvard had been speculating privately by 1834 about the possibility that the course of Uranus was being disturbed by an undiscovered planet. Thomas John Hussey, an English clergyman and amateur astronomer, raised this possibility with Bouvard on a visit to France. When he returned to England, Hussey wrote to George Airy, soon to be astronomer royal, that he and Bouvard independently agreed that there must be an outlying planet. Hussey sought advice about how to find it. Airy discouraged such thoughts, informing Hussey that mathematics was unequal to the task of calculating the position of such a planet.

> It is a puzzling subject, but I give it as my opinion, without hesitation, that it is not yet in such a state as to give the smallest hope of making out the nature of any external action on the planet. . . . But if it were certain that there were any extraneous action, I doubt much the possibility of determining the place of a planet which produced it. I am sure it could not be done till the nature of the irregularity was well determined from several successive revolutions.[10]

In other words, since the revolution of Uranus required 84 years, Airy was telling Hussey to wait several centuries.

In 1835, Halley's Comet arrived at perihelion a day later than predicted. Trying to account for this discrepancy, Benjamin Valz wrote: "I would prefer to have recourse to an invisible planet, located beyond Uranus; its orbit, according to the [Bode's Law] progression of planetary distances . . ." The planet's perturbations on the comet over four or five revolutions might then reveal its location. "Would it not," sighed Valz, "be admirable thus to ascertain the existence of a body which we cannot even observe?"[11]

Friedrich Bernhard Gottfried Nicolai responded to the same stimulus with almost the same reaction. He observed that the delay of the comet in reaching perihelion had been noted previously. "One immediately suspects," wrote Nicolai, "that a trans-Uranian planet (at a radial distance of 38 astronomical units, according to the well-known rule) might be responsible for this phenomenon."[12]

In 1837, Bouvard set his nephew Eugène to work on collecting new position measurements for Uranus to attempt a fresh orbit calculation. Eugène wrote to Airy at the Royal Observatory at Greenwich, explaining that his uncle thought the unruly perturbations were the result of an undiscovered planet. Airy replied that he was confident that the problem could be traced to the adopted distance from Uranus to the Sun. It was too small, he was sure. He urged the young astronomer to concentrate on that possibility instead. Eugène did, and he showed that even with a larger orbit for Uranus, there was still no reconciliation between observation and theory. Nevertheless, Airy continued to hold that the problem lay in the distance from the Sun attributed to Uranus. Thus Airy, who had done much to publicize the discrepancies between the predicted and actual positions of Uranus, was increasingly isolated in his opinion about the cause. By 1838 the existence of a trans-Uranian planet was a major topic in astronomy.

Two astronomers, Niccolò Cacciatore, director of the Palermo Observatory, and Louis François Wartmann, owner of a private observatory in Geneva, thought that independently they might have caught sight of a planet beyond Uranus.[13]

By the end of the 1830s, most astronomers were convinced that an unseen and troublesome planet lurked beyond Uranus. Astronomers such as Johann Heinrich von Mädler expressed "the hope that analysis will at some future time realize in this her highest triumph, a discovery made with the mind's eye, in regions where sight itself was unable to penetrate."[14]

Friedrich Bessel, who for almost 20 years had blamed the trouble with Uranus on an unseen planet, told John Herschel in 1842 that he and his student Friedrich Wilhelm Flemming were preparing to attack the problem, beginning with a new analysis of positional observations of Uranus. But Flemming died suddenly, and soon thereafter Bessel's health failed.

In 1841 a young mathematics student at Cambridge University was browsing in a bookstore and came across Airy's 1832 presentation of the predicament. A week later he wrote himself a note in which he resolved to take on the problem as soon as he finished his degree. This young man was destined to solve the problem and suffer one of the most excruciating frustrations in the history of astronomy. His name was John Couch Adams.

CHAPTER 4

Neptune:

The Planet Found
on a Sheet of Paper

*"(The Uranus problem is) not yet in such a state as to give
the smallest hope of making out the nature of any
external action . . ."*
George Airy discouraging T. J. Hussey in his hope that Neptune could be
found by calculation (1834)

"That star is not on the chart."
Heinrich d'Arrest's response as Johann Galle called off stars in the search
for Neptune (1846)

John Couch Adams was lucky to have survived his early education. His
parents, Thomas and Tabitha, meant well. They were devoted to one
another and their seven children. John was born June 5, 1819, on a farm
near Laneast in Cornwall, England, where his father was a tenant farmer.
The family was "respectable"—not poor, but with very limited means.

John's first school was in a farmhouse in Laneast, but at age eight he
was placed under the tutelage of R. C. Sleep, newly arrived in
southwestern England, who was known to advertise himself as

> Professor of Caligraphy, Stenography, Mathematics, French, Hebrew,
> etc. . . .
> Mr. Sleep Challenges any man in England for Caligraphy, Stenography or
> the Mathematics.[1]

Little did Mr. Sleep reckon on finding a youngster like John Adams
in the provinces. When the boy was ten, student and "professor" sat down
for the first time to study from Mr. Sleep's one and only mathematics

book, an old algebra text. John was almost immediately far ahead of his teacher.

In 1830, at the age of 11, John accompanied his father on a visit to a relative in Devon. The relative noted John's precocity in comparison to his own bright son and arranged a competition between John and the town's schoolmaster, who had a local reputation as a mathematician. It was a humbling experience for the older man.

The next year, the Adamses removed their son from Mr. Sleep's tutelage and sent him off to a school in Devonport headed by Tabitha's cousin. The school was strong in the classics but weak in mathematics. John, age 12, took the matter into his own hands. He spent every possible moment in the nearby town hall, which housed the library of the Devonport Mechanics Institute. There he devoured books and encyclopedia articles on astronomy and, by himself, worked his way through differential calculus.

John Couch Adams at about the time that Neptune was discovered (1846)
National Maritime Museum Greenwich

He was thrilled by the sight of Halley's Comet in 1835 and wrote his first scientific publication for a local newspaper on a lunar eclipse in 1837. Several London papers reprinted the article, much to the delight of the author and the local editor. By now John was reading mathematics texts and devising new solutions to the problems presented. He was seriously preparing himself for college, not only in his studies but also by tutoring the sons of local wealthy families to supplement what little tuition money his family could provide. A local minister who had a Cambridge University education and some competence in math helped him along in his scholastic preparation. In October 1839, John, age 20, left for Cambridge to take the examinations for a sizarship, a scholarship offering financial assistance in exchange for work at St. John's College.

Adams at College

Unlike the stereotypical gifted student, John was modest and gracious. He made friends easily despite the class-conscious and competitive environment at Cambridge. Archibald Samuel Campbell was also there to take the exams and was one of Adams' first acquaintances:

> He and I were the last two at a viva-voce [oral] examination; he broke the ice by asking me to come to his rooms to tea. I went and we naturally had a long talk on mathematics, of which I knew enough to appreciate the great talent of my new friend. I was in despair, for I had gone up to Cambridge with high hopes and now the first man I meet is something infinitely beyond me and whom it was hopeless to think of my beating. If there were many like him, my hopes of success were gone. A few days experiences soon relieved me and I knew what a wonderful man I had met. If I could keep *near* him in the examinations I should do very well.[2]

St. John's College, Cambridge University, as it was when John Couch Adams attended. The first doorway to the right of the tower led to Adams' rooms.
The Master & Fellows of St. John's College, Cambridge

Adams won every scholastic honor available in math as well as first prize in Greek Testament every year he was at Cambridge. During the summers he tutored other students to ease the financial burden on his parents. But he firmly set aside one month each summer to be with his family.

In his student diaries, John sometimes rebukes himself for allowing his interest in astronomy to distract him from other work. In 1840 he and two friends walked from Cambridge to London in 30 hours, in part to visit the Royal Observatory at Greenwich. In the spring of 1841 he at last permitted himself to tour the Cambridge Observatory, with its 11.75-inch (30-centimeter) Northumberland equatorially mounted refracting telescope. He and his friend Campbell soon returned to examine it more closely. A few months later John Adams assigned himself an extraordinary project that set the course of his life and gave that telescope a crucial role in his future.

On June 26, 1841, Adams was browsing in a Cambridge bookshop when he chanced upon an 1832 report by George Airy that mentioned the increasing frustration of astronomers with the errant behavior of Uranus. A week later Adams wrote a note to himself on a slip of paper:

> 1841. July 3. Formed a design, in the beginning of this week, of investigating, as soon as possible after taking my degree, the irregularities in the motion of Uranus, wh[ich] are yet unaccounted for; in order to find whether they may be attributed to the action of an undiscovered planet beyond it; and if possible thence to determine the elements of its orbit, &c approximately, wh[ich] w[oul]d probably lead to its discovery.[3]

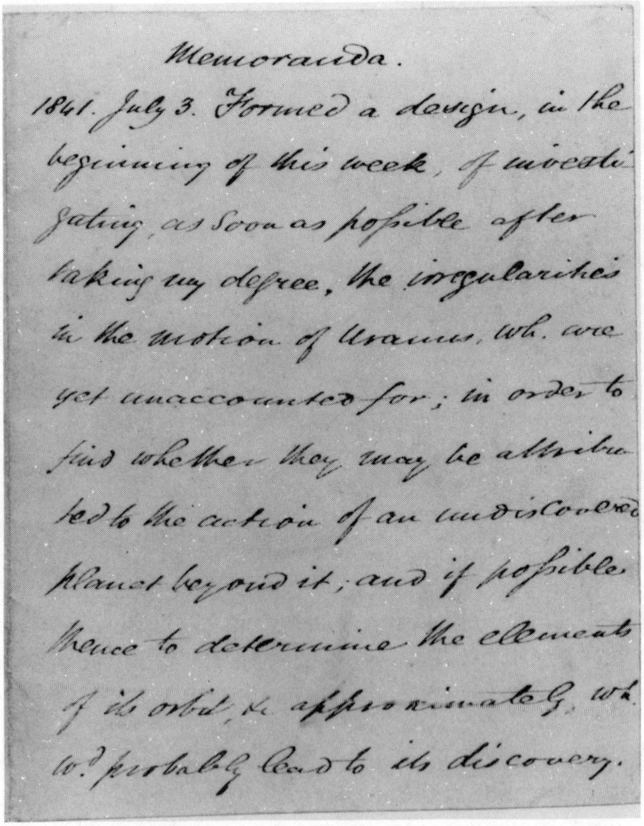

John Couch Adams'
memorandum to
himself on July 3,
1841, in which he
resolved to locate,
using mathematics, an
undiscovered planet
that he thought was
causing irregularities in
the motion of Uranus
*The Master & Fellows of St.
John's College, Cambridge*

Adams drove straight for his objective. He completed his undergraduate degree in 1843 by taking first place in the Mathematical Tripos, a fearsome ordeal of 12 tests, each three hours long, strewn with original problems of extraordinary complexity. Only the top-ranking students, candidates for mathematical honors, were admitted to this examination, which determined postgraduate fellowships and often scientific careers, and occasionally caused nervous breakdowns in excellent students.

Adams emerged as senior wrangler, the foremost student in mathematical science, with a total of 4,000 points. Francis Bashforth, who went on to assume an endowed professorial chair in mathematics, finished second with fewer than 2,000. There was a greater gap in points between Adams and Bashforth than between Bashforth and the worst of the candidates who took the exam. Campbell, by now Adams' close friend, placed fourth. Adams followed this honor by winning the First Smith's Prize, the university's top award in mathematics. He was then elected a fellow of St. John's College.

It was now the spring of 1843, and Adams was anxious to begin work on the Uranus problem. But he felt obligated to his parents because of their sacrifices for his education, so he spent as much time as he could tutoring students and sending the money home to repay his family and to help with the education of his two younger brothers. Only his vacation time was allocated to Uranus. Nevertheless, in the spring of 1843 he outlined his project to James Challis, director of the Cambridge Observatory, and received encouragement.

The problem that John Adams was attacking was so difficult that most mathematicians had written it off as insoluble or, at least, too much of a gamble with their time and careers. Adams didn't know it, but George Airy, the astronomer royal, was already firmly on record that the mathematics could not be done. When noted amateur astronomer Thomas J. Hussey had enthusiastically written to him in 1834 that the perturbations of Uranus pointed to the existence of an outlying planet whose position might be calculated, Airy strongly discouraged the idea, insisting that the motion of Uranus was still much too poorly known to point toward a planet. Even if an outlying planet was disturbing Uranus, he doubted that mathematics could calculate the planet's position. Airy counseled waiting a few centuries until the planet completed several revolutions. And, with characteristic obstinancy, Airy kept his mind closed.

Yet Airy had been senior wrangler and First Smith's prizeman at Cambridge in his college days, then professor of astronomy and director of the observatory there before his appointment as astronomer royal in 1835.

James Challis, who succeeded Airy as Plumian Professor of Astronomy and director of the Cambridge Observatory, had also been senior wrangler and First Smith's prizeman in his time (1825). Yet Challis privately saw no hope in finding a planet based on the motion of Uranus. As he explained later, it was "so novel a thing to undertake observations in reliance on merely theoretical deductions . . ."[4]

Throughout this period, France and Germany produced a succession of noted mathematicians, but no one, until 1845, saw any promise in attacking the problem.

In the standard gravitational problem, it was possible, knowing the existence of Uranus, to calculate the gravitational disturbance it would cause in the motion of Saturn. The relative distances of Saturn and Uranus from the Sun were known. The relative mass of Uranus was known from the revolution of the first of its moons to be discovered. The positions of Uranus and Saturn in their orbits were known from observations and hence the distance and direction from which the gravitational force of Uranus would be exerted. Using these factors to calculate the perturba-

Adams' First Assault

James Challis, director of the Cambridge Observatory
Royal Greenwich Observatory

tion in Saturn's motion caused by Uranus was tedious and required care, but it was straightforward, thanks to Newton.

What Adams had to do was to solve the problem in reverse: Here were irregularities in the motion of Uranus. Subtract the perturbations caused by its moons and by Jupiter and Saturn. Now use the remaining irregularities to deduce exactly where the disturbing force is coming from, how fast and along what path that body is moving, how far away it is, and how massive it is. It was an intimidating project. If the disturbance was coming from a more distant planet, that planet would also be in motion around the Sun and therefore its distance and position with respect to Uranus would be constantly changing. Thus the direction and magnitude of its gravitational force would always be changing too. Nevertheless, calculate the position, orbit, and mass of the disturbing planet, making sure that the unseen object moves along its path so that it causes precisely the perturbations recorded in the motion of Uranus at every point throughout its observed history.

Of course, the farther away the disturbing body is, the more massive it must be to cause those perturbations. But there is no way of knowing the distance or mass of the disturber at the beginning. So start by assuming a distance. Adams did the obvious and picked 38.4 astronomical units, twice the distance of Uranus and in good agreement with the prediction of that mysterious but thus far effective Bode's Law.[5] Next, consider that the undiscovered body does not travel in a perfectly circular orbit, so refine the calculation to take into account the unseen planet's elliptical path. Finally, use the calculated orbit for the new planet to predict where it can be found in the sky for confirmation by telescope.

To handle all the simultaneous variables, the problem required the invention of a new form of mathematical analysis. There weren't many people capable or willing to face that challenge.

By October 1843, at the age of 24, Adams had produced a preliminary solution of the problem that confirmed the existence of a planet beyond Uranus. He shared the results with Challis and asked his help in obtaining the best and latest positional observations of Uranus from the Greenwich Observatory so that by mathematics he could pinpoint where the unseen planet could be found. Challis wrote to Airy, the astronomer royal, on February 3, 1844, to say that Adams was working on the Uranus problem and to request the data Adams needed. Airy responded to Challis with a wealth of information and an invitation for Adams to write him directly.

Adams continued his efforts diligently as time permitted, but he was slowed by his tutorial workload. The Göttingen Academy of Sciences was offering a prize for the best mathematical paper on Uranus and Professor

William Hallowes Miller urged him to enter, but Adams didn't have the time to write up his work in progress.

By the summer of 1844, Adams was working fitfully on his second analysis, which attempted to merge the best current Greenwich observations of Uranus with the old prediscovery plottings and to determine an elliptical rather than a circular orbit for the disturbing planet.

By mid-September 1845 he had completed his new solution to the Uranus problem. He could not know it, of course, but the position he predicted for the unknown planet was accurate to within 2 degrees.

Again he gave his results to Challis, including the predicted position—and from that moment and for the next two years everything that could go wrong with Adams' professional career did.

Challis, who possessed excellent mathematical credentials himself, knew of Adams' reputation for brilliance. Here, confronted with emphatic evidence of Adams' skill, Challis might have immediately undertaken a telescopic search or at least advised the young man to publish his results. He did neither. Instead he suggested that Adams send his results to Airy. Since Adams was about to leave on his monthlong visit home to see his family, Challis provided a letter of introduction so that Adams could deliver his findings to Airy personally.

Adams arrived at the Greenwich Observatory at the end of September and found that Airy had gone to Paris for a meeting of the French In-

Response from Above

Royal Observatory Greenwich about the time that John Adams tried to deliver his calculations for a trans-Uranian planet to George Airy
Royal Greenwich Observatory

stitute on a harbor-improvement project for the port of Cherbourg. So Adams left his letter of introduction and went on to Cornwall. Upon his return from France, Airy wrote Challis that he was sorry to have missed Adams, that he was very much interested in the subject, and that he would welcome correspondence.

Adams returned to Greenwich on October 21, 1845, on his way back to Cambridge. It was early afternoon, and Airy was in London at a meeting on the standardization of railway track gauges for Britain. But he was expected back soon. Adams left his card and a short statement of his results, saying that he would return later. The card was probably delivered to Mrs. Airy, who may have forgotten to pass the message to her husband. Richarda Airy could have had a lot on her mind. A week later, she gave birth to their son Osmund.[6]

Airy returned soon after Adams left, and at precisely 3:30 in the afternoon he and his wife sat down to dinner. It was Airy's unique and unpublicized custom to have dinner in the middle of the afternoon because his doctor had prescribed this ritual for his supposedly frail patient (who lived to age 90).

Adams returned an hour later, as promised, with the hope of meeting the astronomer royal at last and discussing what he had accomplished. He was met at the door by the butler. Dr. Airy was at dinner and could not be disturbed. No, the astronomer royal had left no message for the young man. What could Adams think? Airy knew his name from correspondence with Challis. Airy had his letter of introduction from Challis. Airy had the summary of his research. The visitor had timed his return to be far from the traditional dinnertime or even teatime. Adams left deeply disappointed.[7]

The astronomer royal may have been unaware of Adams' treatment in his three thwarted visits, but he was in possession of Adams' results. The astronomer royal was unimpressed. His attitude toward Adams was probably compounded of two factors: his dislike of the theoretical approach to scientific research and his snobbishness.

Despite his considerable mathematical abilities, Airy thought that astronomy and all science should proceed by collecting observations and then using mathematics as the language to express the findings. The procedure should not be reversed; mathematics should not be used to point toward observational breakthroughs. Such a theoretical approach was just playing with numbers. A mental game with no basis in reality. A waste of time. He distrusted and even intensely resented such efforts.

Airy's unresponsiveness to Adams was also a product of the snobbishness he had cultivated from his youth. In his view there were two kinds of people in the world: those who had succeeded and those who had not. Successful people were potentially of use and therefore worth cultivating. Those who had not succeeded or had not yet succeeded (such as younger people) were of little use and therefore not worthy of consideration.[8]

Nevertheless, one week later, Airy showed Adams' note summarizing his results to William Rutter Dawes, a clergyman, physician, and extremely skillful amateur observer. Dawes was impressed, but perhaps because he was erecting and furnishing his new observatory, he did not undertake a search himself. He did, however, pass the information and the planet's predicted position along to William Lassell, a successful brewer in Liverpool and the builder of a 24-inch reflector—the largest operational telescope at that time in England. Dawes recommended that Lassell take on the project. Lassell would have liked the challenge, but he was confined to bed with a severely sprained ankle. Before he could return to observing, the letter had disappeared and Lassell lost the opportunity.[9]

Dawes' very positive reaction to Adams' work impressed Airy not at all. The astronomer royal, compulsively orderly and prompt, who always answered his mail immediately, remained so negative to Adams' work that he delayed his response to Adams' paper. On November 5, 1845, more than two weeks after Adams had been turned away from his door, Airy finally replied to the young man with a rejection in the form of a question. Referring to Adams' calculations as assumptions, Airy asked whether Adams' work could explain why Uranus was slightly farther from the Sun than previously calculated. From Adams' point of view, the question was trivial and besides the point. Adams' research could specify the location of the planet disturbing the motion of Uranus. Recent observational data had shown that Uranus was moving more slowly than expected. According to Kepler's third law, the slower a known planet moves, the farther it must be from the Sun. But the problem at hand was the *location* of the *planet disturbing Uranus.* The observed motion of Uranus, whether slow or fast, formed the data from which the location of the disturbing planet was calculated—and this was what Adams had accomplished. Airy's question was essentially irrelevant.

What was apparent was that Airy had completely and inexplicably misunderstood the essence of Adams' work. Adams had *calculated* the planet's existence from royal observatory data that Airy had furnished. He had not *assumed* that the planet existed and guessed at a possible orbit. Airy knew this from the letter of introduction that Challis wrote for Adams, which emphasized that the young man had calculated a planet beyond Uranus using observational data that Airy himself had furnished. The letter also acknowledged Adams as a very competent mathematician. Airy must have read Adams' summary in a most cursory manner.

As he looked at Airy's reply, Adams saw no flicker of recognition or interest in his project, which the astronomer royal had known about for a year and a half. There was no offer to undertake a search. There was no suggestion that his work was worth publishing. With no interest in a search expressed by Challis or Airy, Adams was uncertain how to proceed. He didn't respond immediately to Airy's inane question.[10] Instead, he resolved to rework his calculations to get a prediction so impressive that it could not be ignored.

Adams' failure to reply immediately gave Airy just the excuse he sought to drop the matter: "Adams' silence . . . was so far unfortunate that it interposed an effectual barrier to all further communication. It was clearly impossible for me to write to him again."[11]

On the Continent

The erratic motion of Uranus bothered French, German, and other European astronomers no less than the English. The dean of French astronomers was Jean Dominique François Arago. He was popular with

GEORGE BIDDELL AIRY: ASTRONOMER ROYAL

Young George was not popular with his classmates at school because, he said, he had very little "animal vitality." He was also snobbish and conceited. But his classmates endured him when he created useful gadgets for mischief such as peashooters.

At the age of 12, George, already concerned for his own advancement, conspired with his mother's brother, a wealthy and well-educated farmer, to adopt and hide him from his family without informing them so that he would have better social contacts. His father, depressed over the loss of his job with the tax office, declined to block the arrangement.

At the age of 18, Airy was off to Cambridge University with a scholarship and a very high opinion of himself. He made no friends, and professors judged his abilities to be limited, but Airy worked doggedly and carried off the two big science and mathematics prizes as senior wrangler and First Smith's prizeman in 1823. He thought well enough of himself by this time to write his first autobiography.

Just as meticulously as he recorded his younger days, so he carefully plotted his future life. He kept a complete daily record of his activities and thoughts. His financial accounts were personally kept by double entry throughout his life, and he regarded their keeping as one of his greatest joys.[1] He never destroyed a document and preserved all his old checkbook stubs, bills, and receipts from merchants in chronological order. A colleague at Cambridge quipped, "if Airy wiped his pen on a piece of blotting-paper, he would duly endorse the blotting-paper with the date and particulars of its use, and file it away amongst his papers."[2]

As a fellow of Trinity College in 1824, he set his sights on the position of astronomer royal and refused to accept an assistantship at the Royal Observatory because assistants had not previously been promoted to the post. Instead he fought his way upward through professorships at Cambridge, campaigning successfully for higher pay. In 1835

[1]George Biddell Airy, *Autobiography*, edited by Wilfrid Airy (Cambridge: At the University Press, 1896), p. 2.

[2]This story, told by Augustus De Morgan, appears in E. Walter Maunder, *The Royal Observatory Greenwich* (London: Religious Tract Society, 1900), pp. 116–117.

the public because of his clear, enthusiastic, and engaging way of explaining science. He was popular with astronomers because he liked to identify interesting scientific problems and pass them along to young astronomers for solution. He would then encourage and assist them with his insight and his encyclopedic knowledge of the scientific literature. He also helped to find them jobs.

In the summer of 1845, Arago urged Urbain Jean Joseph Le Verrier to take on the problem of the motion of Uranus. Arago was a shrewd judge of talent who could pair the scientific difficulty of a problem with the intellectual capability of a scientist.

Le Verrier, like Adams, had come from a family of modest means in the provinces. He was born in Saint-Lô in Normandy on March 11, 1811. His father, an estate manager, did all he could to foster his son's intellectual pursuits, and Urbain responded by reaching the top of his class at high school in Caen. He then took the highly competitive exam for the École Polytechnique in Paris, but there were gaps in his provincial education

George Biddell Airy, astronomer royal
National Maritime Museum Greenwich

he received the call he expected as the seventh astronomer royal. He immediately set to work restructuring the Royal Observatory in his mold. Airy was an organizer rather than a scientist.[3]

Under his direction the Royal Observatory was a superb servant of the Navy, providing tables for celestial navigation of improved precision. Greenwich expanded into meteorological and magnetic record keeping. But among staff members, Airy tolerated no independent thought or research. Observers were often required to work 21 hours straight. Young boys hired to do arithmetic calculations were kept at their desks for 12-hour shifts with no breaks except for one hour at midday. Morale was poor. But Airy's efficiency and discipline at the Royal Observatory were admired in England and copied by many other countries, which led years later, particularly in England, to a conspicuous lack of creativity in observational astronomy.

Airy declined knighthood three times, complaining about the initiation fees and the cost of maintaining the expected life-style, but finally accepted the honor in 1872. He retired as astronomer royal in 1881 at the age of 80 and spent his remaining ten years filing (but not organizing the contents of) his papers.

[3]Olin J. Eggen, "George Biddell Airy," *Dictionary of Scientific Biography* (New York: Charles Scribner's Sons, 1974).

and he was not admitted. Urbain's father thereupon sold the family house for cash to enroll his son in the Collège de Saint Louis in Paris for additional preparation in 1830. Urbain rewarded his father's sacrifices by winning the annual mathematics prize sponsored by the École Polytechnique in 1831 and gaining admission to the school. He was never again below the first rank of scientists. He graduated with highest honors, a wide range of scientific interests, and a reputation for tenacity.

His first job was with the Ministry of Tobacco. He had decided on a career in chemistry, and his objective was to work with renowned chemist Joseph Louis Gay-Lussac. Le Verrier fell so in love with Paris and his future wife that he left government service two years later rather than accept the usual tour of duty in the provinces. He spent a year teaching and then rejoined Gay-Lussac when his mentor was appointed to the École Polytechnique faculty and created an assistant's position for him.

Le Verrier's first published papers were in chemistry. Gay-Lussac saw talent there, but even greater skills in mathematics. When a professorship in astronomy opened at the École Polytechnique in 1837, he arranged for Le Verrier to be appointed and encouraged his colleague, now 26 years old, to switch fields. Le Verrier did, studying astronomy with his characteristic ferocity and applying his mathematical incisiveness to it. He worked extremely hard so that his initial publications in astronomy would be worthy of his distinguished predecessor Félix Savary, of his mentor Gay-Lussac, and of his own high standards. Le Verrier's first two papers showed extraordinary analytical powers, and he was befriended (as most promising young scientists were) by Arago in 1840.

Arago gave him the problem of the motion of the planet Mercury, which was always ahead of where its calculated orbit predicted it would be. Le Verrier worked on the problem for three years and explained most of the variance as perturbations caused by other planets. But he was left with a small discrepancy that he could not explain.

With the Mercury problem initially intractable, Le Verrier turned his attention to cometary orbits with great success. In June 1845, Arago approached him again—this time with the most difficult and intriguing puzzle facing contemporary astronomers: the motion of Uranus. Le Verrier was now 34, eight years older than Adams. He was already internationally known for his mathematical analysis of astronomical problems.

Le Verrier began with an exhaustive analysis of the motion of Uranus as presented in Bouvard's *Tables*. On November 10, 1845, Le Verrier presented to the Paris Academy of Sciences the first of three papers on Uranus. It dealt exclusively with old and modern observations and demonstrated that they were irreconcilable—there really was a problem with the planet's motion, not with the quality of the observations. He

François Arago, director of the Paris Observatory
Royal Greenwich Observatory

would, he announced, analyze those outside causes in a second paper soon. He also pointed out that Bouvard was wrong in ignoring the positions of Uranus observed prior to its recognition as a planet.

On November 10, 1845, while Le Verrier was giving his preliminary report on the project that Arago had encouraged him to undertake, Adams was receiving his astronomer royal's most discouraging response to his calculation of where a new planet could be found.

The insult, intentional or careless, was compounded a few weeks later when Airy visited Challis at Cambridge University. Neither Challis nor Airy invited Adams to meet Airy at last and discuss his work with him.

Meanwhile, in France, Le Verrier was hard at work preparing his second paper on Uranus. He presented it to the Paris Academy of Sciences on June 1, 1846. This paper demonstrated that the irregularities in the motion of Uranus could be induced only by an unknown planet farther from the Sun. What had been suspected now had a solid mathematical basis that had been announced to a wide scientific community.

Le Verrier then went on to calculate the planet's position, ending with the hope that "we will succeed in sighting the planet whose position I have just given."[12] The academy praised this new analysis, but to Le Verrier's amazement, no one offered or even suggested that a search should be undertaken.

Airy received a copy of Le Verrier's second paper on June 23 or 24, 1846. He immediately noticed that Le Verrier's prediction agreed to within one degree of Adams'. "To this time," wrote Airy, "I had considered that there was still room for doubt of the accuracy of Mr. Adams's investigations . . . But now I felt no doubt of the accuracy of both calculations . . ."[13]

Airy responded immediately with a letter to Le Verrier expressing enthusiasm and delight. Le Verrier was eight years older than Adams and an established scientist. He was worthy of notice. Airy's response was also in character, wrongheaded and neglectful. Had Le Verrier considered the possible need to modify the length of the Uranus radius vector (the Sun-to-Uranus distance)? It was the same inane question he had asked Adams. But Airy didn't bother to mention that Le Verrier's fine work had been matched eight months earlier by a young mathematician at Cambridge. And, of course, Airy did not bother to share his "delight and satisfaction" with Adams.

Airy Begins to Move

Airy must have been even more impressed than his reply to Le Verrier indicated, because three days later, when the Board of Visitors (the commission that conducted periodic inspections) of the Royal Observatory convened at Greenwich on June 29, Airy proclaimed "the extreme probability of now discovering a new planet in a very short time . . ." He was confident because Adams' and Le Verrier's calculations of the position of the planet disturbing Uranus coincided so closely. Still Airy made no effort to notify Adams, nor did he bother to authorize a search. Serving on this board, hearing these words, were James Challis and John Herschel.[14]

Upon hearing this news, Challis behaved inexplicably. Instead of finally recognizing the opportunity that had been staring him in the face for a year, instead of starting a search on his own, Challis now lost whatever little interest he had in the matter and what little confidence he had in Adams' predictions. Perhaps he thought that Airy had preempted the project.

Le Verrier received Airy's reply on June 28. Since he had dealt with Airy only a few times previously, he was impatient rather than discouraged. He immediately informed Airy in firm icy tones that the astronomer royal's question was irrelevant because his calculations dealt with perturbations on Uranus and where they were coming from, hence the exact distance of Uranus from the Sun was corrected for "automatically." Having dismissed the astronomer royal's objection, Le Verrier went on to offer to send specially calculated position predictions to simplify the quest for the new planet if Airy had "enough confidence in my work to search for the planet in the sky . . ." Airy received the offer on July 1 and again responded immediately. He declined Le Verrier's offer because, he said, he was about to leave for business on the continent. His trip was actually scheduled for August 10, five and a half weeks away.

The next day Airy again visited Cambridge University. He was in the company of visiting astronomer Peter Andreas Hansen, director of the Seeberg Observatory near Gotha, Germany (which Zach had helped to build and had directed). Despite Le Verrier's recent letter and despite his own statement to his observatory's Board of Visitors that the irregular motion of Uranus would soon lead to the discovery of a new planet, the astronomer royal made no effort to contact or meet with Adams. It happened anyway, but accidentally. That evening Airy and Hansen encountered Adams on St. John's Bridge. Airy introduced Adams to Hansen but was very cool and made no mention of Adams' calculations for a new planet. He cut off the conversation after about two minutes and moved on.

A few days later, on July 6, Airy spent some time at Ely with George Peacock, his old professor at Cambridge, and the subject of Uranus was broached. Airy told his mentor about his correspondence. Peacock was dumbfounded by his former student's negligence and urged him to act.

Airy finally got the message, and three days later, on July 9, he wrote James Challis to request an urgent search. Presuming Challis' answer would be negative, because such a project would likely be time-consuming and tedious, Airy offered an assistant from the Royal Observatory to help. The situation, said Airy, was "almost desperate," neglecting to acknowledge that he had delayed a search for nine months.

Four days later, without waiting for Challis to reply, Airy wrote again, this time sending him a complete set of instructions for the search. Nothing except emergency unpostponable observations should be allowed to intervene, said Airy, because "the importance of this inquiry exceeds that of any current work . . ."

Challis was away from Cambridge when the letters arrived. When he returned, he was of course offended by Airy's imperious and condescending manner. In his reply of July 18, Challis told Airy that he had already decided to look for the planet himself and therefore declined an assistant.

It made sense for the Cambridge Observatory, rather than the Greenwich Observatory, to conduct the search. Greenwich primarily had transit telescopes, which could pivot up and down but not side to side. They were locked into position along the meridian so that they could observe exactly when a star, planet, the Moon, or the Sun crossed the north-south line for purposes of calculating time, longitude, and orbits. The Cambridge Observatory had the 11.75-inch (30-centimeter) Northumberland Telescope, equatorially mounted so that it could be aimed at any point in the sky and with more than enough light-gathering power for this kind of search.

Challis Begins to Act

But for the next week and a half Challis did nothing. Then, late in July, Challis told Adams of the projected search. Adams provided updated position predictions for the planet and gave Challis very encouraging news: The planet should be large enough to show a disk. Thus a careful observer would be able to recognize the new planet as other than a star without having to wait a day or so to see if the suspected object moved among the background stars.

On July 29, 1846, Challis began his search, resigned to the prospect of a long, tedious, and fruitless task because mathematics was no way to find a planet. Placing no confidence in the quality of Adams' or Le Verrier's calculations and ignoring Adams' advice that the planet should show a disk, Challis agreed with Airy that it was useless to concentrate on the positions that Adams and Le Verrier had predicted. Instead he prepared to search a swath of sky 30 degrees long and 10 degrees wide— 15 degrees to the east and west and 5 degrees to the north and south of the predictions. Challis' plan was to sweep this region steadily, record-ing all but the dimmest stars, and then to return to that region a few days later to reexamine each star to see if one had moved. Such a search would require mapping at least 3,000 stars and would consume at least 300 hours of telescope time—approximately eight straight weeks of good observ-ing weather. This was the only way, unless he was to value the work of Adams and Le Verrier and concentrate his search on the region of the sky where they predicted the planet would be. If he had, his efforts would have succeeded within a few nights.

On August 12, 1846, two weeks into his search, but on only his fourth night of acceptable observing conditions, Challis and his assistants re-corded in his notebook the new planet almost exactly where it was predicted. But Challis failed to recognize his target.

Le Verrier's second paper, identifying an outlying planet as the source of Uranus' peculiar motion and predicting its position, circulated wide-ly. Sears Cook Walker, a young astronomer at the United States Naval Observatory in Washington, D.C., read of Le Verrier's work and proposed to search. But Matthew Fontaine Maury, superintendent of the obser-vatory, rejected the proposal because of his facility's heavy schedule.

Final Predictions

Le Verrier kept pushing forward. On August 31, 1846, he presented his third and final paper on Uranus to the Paris Academy of Sciences, this time laying out the orbital elements, the mass, and the position of the planet that was disturbing Uranus. His refined positional calculation dif-fered only slightly from his previous prediction. The planet, he said, was about 5 degrees east of Delta Capricorni. To encourage a search, Le Ver-rier stressed that the planet was just past opposition—closest to Earth and visible almost all night long. The planet was big enough to show a disk, so that a long, tedious mapping of stars would not be necessary. He received polite applause from his scientific audience and, once again, not a single offer to search for the planet from any astronomer, not even his champion Arago.

Adams did not know of Le Verrier's third paper when, on September 2, he sent Airy his sixth solution to the problem, refining a little bit fur-ther his prediction of where the disturbing planet would be found. This calculation explicitly included a correction for the distance of Uranus from the Sun, the matter about which Airy had previously expressed con-cern. Airy was now actually away on his trip to the continent, and Adams received a mindless reply from Robert Main, Airy's chief assistant, of-fering more observational data on Uranus.

Adams now at last resolved to lay his work before the scientific community. He decided to present his findings at a meeting of the British Association for the Advancement of Science in Southampton in mid September. But when he arrived on September 15, he found that there had been a confusion in the conference announcement. The astronomy session had already concluded, and there was no other appropriate forum for him. The irony was overwhelming when Adams heard that John Herschel had opened the conference with his valedictory address as president of the association by heralding the prospect of a new planet: "Its movements have been felt . . . with a certainty hardly inferior to that of ocular demonstration."[15]

Johann Galle

In France, Le Verrier was also frustrated. His countrymen praised his mathematical virtuosity, but no one was willing to invest time in a search for the planet. Airy in England had declined to search, as far as he knew. Then Le Verrier remembered a doctoral dissertation he had received from Johann Gottfried Galle, an assistant at the Berlin Observatory. Almost a year had passed and Le Verrier had failed to acknowledge the gift. On September 18, 1846, Le Verrier picked up his pen, praised the young astronomer, and enclosed his position predictions for the new planet with the request that Galle attempt a search.

Berlin

The letter reached Galle on September 23, 1846. He rushed into the office of Johann Franz Encke, the director of the Berlin Observatory, to request permission to use the institution's fine 9-inch (23-centimeter) refracting telescope for the project. It took considerable pleading before Encke gave way. It was Encke's birthday and he intended to celebrate at home with his family, so he had no personal plans for the telescope that night.[16] Heinrich d'Arrest, a young graduate student at the observatory, overheard the discussion in progress, boldly joined in, and begged to be allowed to help. "Let us oblige the gentlemen in Paris," said Encke finally, and departed.[17]

Galle and d'Arrest could scarcely wait for nightfall. They opened up the observatory dome and started looking for the new planet. As d'Arrest stood by, Galle searched the region around the position specified by Le Verrier—right ascension 21 hours 46 minutes, declination –13 degrees 24 minutes—in hopes of identifying the planet by its disk. But the planet was not immediately evident in the field of view and in surrounding fields.

Johann Encke, director of the Berlin Observatory

Then d'Arrest suggested that they use a star map and compare stars in the sky with those on the chart to find one that wasn't plotted. Galle was reluctant. The star maps he had been using (such as those by Carl Ludwig Harding) were not very reliable. Nevertheless, there seemed to be no other choice.

The Berlin Observatory
as it was about the
time that Neptune was
discovered

Galle and d'Arrest hunted through the map files. There, to their delight, they found an excellent new chart of the region in Aquarius that they needed. They had not known that this particular map existed. Prepared by Carl Bremiker, their own observatory's staff mathematician and map-maker, this chart was a product of the Berlin Observatory's involvement with other German observatories in a full-sky mapping project inaugurated in 1830 at the urging of Friedrich Wilhelm Bessel to help with the search for more asteroids. (No new minor planets had been discovered since 1807.) Bremiker's chart of Hora XXI (right ascension hour 21 and one hour to either side) had been printed earlier that year but had not yet been released for distribution, possibly so that the Berlin Observatory could enjoy a temporary advantage.[18]

The two astronomers went back to work: Galle with his eye to the telescope, calling out each star in his field of view; d'Arrest off in a corner with a dim lamp shielded from Galle's eyes, matching each star as it was announced with one of the right position and brightness on the map.

"On the chart," he would respond. Galle would call another.

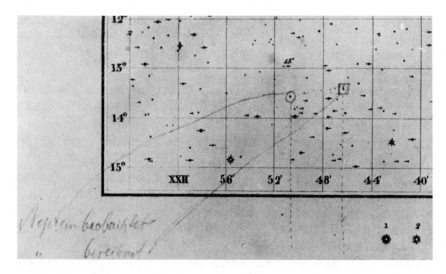

Star map by Carl Bremiker that Johann Galle and Heinrich d'Arrest used to find Neptune. The position circled is marked *Neptun beobachtet* ("Neptune observed"). The position with a square around it is labeled *Neptun berechnet* ("Neptune calculated").
Archive of the Archenhold Observatory

"On the chart," d'Arrest would say again.

Galle moved the telescope slightly to widen the search by one degree and resumed calling.

"On the chart," came the answer.

"Right ascension 21 hours 53 minutes 25.84 seconds; magnitude 8," said Galle.

"That star is not on the chart," said d'Arrest.

They had examined only a few stars. It was under an hour since they started observing. The object was less than one degree from where Le Verrier had predicted.[19]

They ran to find Encke, breaking up the birthday party. The three of them observed until the early morning hours when the object set, but they could not yet be sure that it was moving against the background stars, nor could they be sure that it showed a disk.

Encke joined Galle and d'Arrest for the commencement of observing the next night, September 24, 1846. The observing conditions were even better than the night before. The first look told the story. The object had moved. Moveover, it displayed a tiny disk, just as Le Verrier had predicted.

Galle sent notification to Le Verrier the next morning, September 25, beginning exuberantly, "The planet whose position you have pointed out *actually exists.*"[20] Concluding his praise for Le Verrier, Galle suggested that the new planet be named Janus.

Meanwhile, Encke began notifying the German astronomical community, with credit for finding the planet to Galle and himself, while omitting acknowledgment for the graduate student d'Arrest.

The Berlin Observatory's 9-inch refracting telescope with which Johann Galle and Heinrich d'Arrest found Neptune, based on the calculated position furnished by Le Verrier. The telescope is now on display at the Deutsches Museum in Munich.

Meanwhile . . .

News of the planet's discovery reached Airy while he was still on the continent, returning the visit of Peter Andreas Hansen at the Seeberg Observatory he directed near Gotha. It was September 29, two days before the news reached England. Airy's reaction was not recorded.

Meanwhile, in England, James Challis was plodding along with the search assigned to him. On September 29, 1846, he finally received a copy of Le Verrier's third paper on Uranus, giving the position of the unknown planet and the prediction that it would exhibit a disk. Challis had ignored that suggestion when it came from Adams but now he began looking for the disk. That very night, he found an object with a disk and noted its position, but didn't bother to follow up—to examine the object at higher magnification to be sure. Perhaps the next evening.

At dinner on September 30,[21] Challis mentioned the search and the disk he thought he saw to the Reverend William Towler Kingsley of Sidney Sussex College. Kingsley surprised Challis with his great excitement and asked to see it for himself at a higher power. Challis proposed a look after dinner. They reached the observatory building, which also served as the Challis home, under clear skies. When they arrived, Mrs. Challis insisted on serving them tea. By the time they finished their refreshment, the sky had clouded over. The days that followed, Challis felt, would find the Moon interfering with observations in that part of the sky, so no observations were attempted.

On October 1, 1846, the *Times* of London announced the discovery of the new planet, confirmed by John Russell Hind, an observer in London, who found it easily with a 7-inch (18-centimeter) telescope despite the moonlight and hazy skies.

Challis reexamined his records and found that he had seen the new planet twice in his first four days of searching but had not recognized it. In a letter to Airy, he confessed: "after four days of observing, the planet was in my grasp, if only I had examined or mapped the observations."[22] He had seen it again as a disk on September 29.

France

In Paris, Le Verrier was soaking up the adulation and pondering a name for the planet. On October 1 he wrote to Galle: "I thank you cordially for the alacrity with which you applied my instructions to your observations of September 23rd and 24th. We are thereby, thanks to you, definitely in possession of a new world."[23]

But Le Verrier was annoyed by Galle's suggestion of the name Janus. That was trespassing on the discoverer's right to name the planet. So he appended a postscript to the letter stating that the French Bureau of

Longitudes had named the planet Neptune. It was not true. The idea was Le Verrier's. He may have mentioned it to friends of his at the bureau who unofficially thought the name a good one, but the bureau had no official role in assigning astronomical names. Le Verrier was content that he had headed off Galle's proposal with one that seemed to carry the weight of authority.

But during the next few days Le Verrier had a change of heart. He decided that the planet should be named after himself. Yet he had already sent out notifications of his previous choice of the name Neptune to a number of astronomers. How could he undo this damage?

He went to see François Arago, the dean of French science, his friend, and the man who had sent him on the quest for this suspected planet. Arago was reluctant at first. He thought Neptune was a sensible name. But he was so impressed with Le Verrier's work that he agreed, on one condition: He would propose and urge the adoption of Le Verrier as the planet's name if the name of Uranus could be changed to Herschel (the name the French had originally proposed when Uranus was discovered).

It was a shrewd political approach. It was also very awkward and unseemly. Le Verrier's collected papers on Uranus, nearing publication, received a new title: *Recherches sur le mouvement de la planète Herschel (dite Uranus)* (Research on the Motion of the Planet Herschel [called Uranus]), but the text of the article was too far along for alteration and used the name Uranus exclusively. Le Verrier appended a note: "In my future researches, I shall consider it my strict duty to eliminate the name Uranus completely, and to call the planet only by the name *Herschel*."

On October 5, Arago announced to the Paris Academy of Sciences that Le Verrier had invited him to name the new planet and that he had chosen the name Le Verrier. He made the best possible case for this choice in an emotional speech. Comets were named after their discoverers. Why not planets, which are much greater discoveries and particularly this planet, found by such "an admirable and unprecedented method."[24] So the planet Uranus ought to be called Herschel, and the asteroid Juno ought to be Olbers (to satisfy English and German pride). Such names, he argued, represented legitimate patriotism.

The idea was harshly rejected by the international astronomical community, which had already informally adopted Le Verrier's original suggestion of Neptune. For the moment, the most serious problem had become the name for the planet.

Le Verrier and his admirers strutted their victory for the glory of France. Letters poured in. H. C. Schumacher's comment on Le Verrier's work was typical: "It is the most noble triumph of theory which I know of."[25]

The English Claim

John Herschel
Courtesy of Special Collections, San Diego State University Library

Little did the French expect that the English were about to rain on their parade. John Herschel wrote to set the record straight and give Adams appropriate credit. Airy and Challis wrote to exonerate themselves and to stake out their places in history.

On October 3 an article by John Herschel appeared in the London weekly magazine *Athenaeum*. After congratulating Le Verrier, Herschel quoted his speech three weeks earlier at the meeting of the British Association for the Advancement of Science in which he forecast that the predicted planet would soon be found. Just one such calculation would not have given him the confidence to make that prediction, he explained. He had heard from Airy that Adams had made a similar calculation earlier:

> But it was known to me, at that time, (I will take the liberty to cite the Astronomer-Royal as my authority) that a similar investigation had been independently entered into, and a conclusion as to the situation of a new planet very nearly coincident with M. Le Verrier's arrived at (in entire ignorance of his conclusions), by a young Cambridge mathematician, Mr. Adams—who will, I hope, pardon this mention of his name (the matter being one of great historical moment)—and who will, doubtless, in his own good time and manner, place his calculations before the public.

On October 5, Challis wrote to Arago to say that he had been searching for the new planet since July 29 and that, upon receiving Le Verrier's third paper on September 29, he searched for and found that night a star with a disk—prior to his hearing about the Berlin Observatory discovery six days earlier. Challis was trying to grab a bit of the fame for himself.

What Challis omitted from his letter says much about his character. He did not mention that the search that he began on July 29 was based on Adams' computations. In fact, Challis made no mention of Adams and his work at all. Challis also neglected to admit that he had seen the planet twice previous to its identification at the Berlin Observatory but had not bothered to check his work and therefore failed to recognize the planet. He also failed to mention that he had not bothered to positively identify the new planet when he saw it on September 29.

While the astronomers involved were congratulating or positioning themselves, other observers were examining the new object. William Lassell, the brewer and amateur astronomer who might have discovered Neptune a year earlier if he had not been incapacitated by an ankle injury and if his copy of Adams' position predictions had not been lost, went to work examining Neptune with his 24-inch (61-centimeter) reflector. On October 10, 1846, two and a half weeks after Neptune was found, Lassell discovered that Neptune had a moon. But before this moon could be positively confirmed (there weren't many other telescopes capable of seeing so faint an object), the Sun intruded too close to the planet's position in the sky for favorable observation. The moon of Neptune, now called Triton, was not confirmed until July 1847.

On October 14, 1846, it was Airy's turn to stake out his position. He realized he was trapped and was trying to shape the circumstances to his best advantage. He began his letter to Le Verrier with effusive congratulations. Then came the bombshell:

Airy Maneuvers

> I do not know whether you are aware that collateral researches had been going on in England and that they led to precisely the same results as yours.

Of course, Le Verrier didn't know of Adams' work. During his correspondence with Le Verrier on Uranus and its unseen disturber, Airy had deliberately withheld mention of Adams. Then Airy tried to have his cake and eat it too, and in so doing, he committed another inexcusable denigration of Adams' work.

> I think it probable that I shall be called on to give an account of these. If in this I shall give praise to others, I beg you will not consider it as at all interfering with my acknowledgements of your claims. You are to be recognized beyond doubt as the real predictor of the planet's place. I may add that the English investigations, as I believe, were not quite so extensive as yours. They were known to me earlier than yours..[26]

Even so, Airy, in his letter, still could not bring himself to mention Adams by name.

Finished with that, Airy continued his maneuvers. He answered a letter from Challis in which the Cambridge professor reported on his failed search and then suggested the name Oceanus for the new planet. Airy pretended to be unconcerned about the failure—"these misses are sometimes nearly unavoidable"—and then went on to recommend that Challis present the name Oceanus to Le Verrier, even though Airy had privately and in correspondence lent the weight of his opinion to the name Neptune and even though he knew that Le Verrier already had his own name in mind for the planet. It was a rather shabby diversionary tactic. Airy could see the scientific and political storm gathering. Having lulled Challis into a sense that no harm was done, Airy asked Challis for permission to publish his report on the planet search. Airy then wrote to Adams to make the same request so that England and "individuals" would receive "justice." Airy proposed to prepare this report himself for the Royal Astronomical Society "because I know nearly all the history and yet have taken no part in the theory or the observations."[27] Airy, who knew "nearly all the history," sent this request to the "Rev. W. J. Adams." Adams' initials, of course, were J. C., and he was not ordained.

Airy's letter reached Le Verrier on October 16. About the same time he learned of Herschel's article in the *Athenaeum*. Le Verrier was furious. If this Adams had done the work, why had neither Airy nor Challis ever mentioned him? Who was this Adams? He was professionally unknown. If he had done the work, why had it not been published? And why had nothing been heard from Adams himself? Le Verrier wrote to Airy to pose these embarrassing questions and to ask the astronomer royal to defend his claims.

The next day, on October 17, Challis published an account of his investigation in the *Athenaeum*, mentioning Adams' first solution of the Uranus problem in 1843. He also proposed the name Oceanus for the new planet.

Resentments

That did it. On October 19, 1846, the Paris Academy of Sciences met and fumed like a hall of politicians over what seemed like a British effort to usurp their nation's honor. Arago used his gift of rhetoric to discredit the English claims and to inflame the academy members. Arago was surprised by Herschel's remarks, "in complete contrast with the usual courtesy and reserve of Mr. Herschel."[28] But he reserved his venom for Airy and Challis. The astronomer royal was an authority on celestial mechanics. It was impossible to believe, if Airy had really seen Adams' work, that he could have doubted the existence of a trans-Uranian planet.

As for Challis, he was doomed by the inconsistencies in the reports he had filed. In his October 5 letter to Arago, Challis had mentioned only Le Verrier as a spur for his search. But in the October 17 *Athenaeum* article, Challis claimed to have been guided by Adams' work. This discrepancy made Challis appear either dishonest or deranged.

Since Adams had not published, said Arago, he had no claim. Concerning Oceanus as a name for the planet, which Arago thought had come from Adams, it was the height of arrogance for someone who had contributed nothing to try to horn in on the glory. "Mr. Adams has no right to figure in the history of the discovery of the planet Le Verrier, neither by a detailed citation, nor by the slightest allusion." This discovery, concluded Arago, "one of the most magnificent triumphs of astronomical theory," was a contribution to posterity from France.[29]

The French press picked up the issue and the tone and began a series of vicious attacks on Adams, Airy, Challis, Herschel, and England generally. It got so ugly that Arago and Le Verrier soon disavowed their sympathy with the papers.

While the dispute was raging, the influential scientists on the continent quietly, through correspondence, settled on the name Neptune for the planet and repudiated the use of Le Verrier's name.

The Investigation

On November 13, 1846, the Royal Astronomical Society held an inquiry into what now appeared as the Neptune scandal. How had the English scientific establishment let this prize slip through their fingers? Airy presented his "Account of Some Circumstances Historically Connected with the Discovery of the Planet Exterior to Uranus." Challis followed with his "Account of Observations at the Cambridge Observatory for Detecting the Planet Exterior to Uranus." Then Adams offered "An Explanation of the Observed Irregularities in the Motion of *Uranus* . . . ," complete with the position, mass, and orbit of Neptune—presenting an abstract of the work that he had tried in vain for more than a year to bring to the attention of the scientific establishment.[30] Adams had also computed the first orbit for Neptune, using the three unrecognized sightings by Challis, and had sent it to Airy on October 15.

Adams expressed no disappointment or bitterness, only praise for Le Verrier:

> I mention these dates merely to shew that my results were arrived at independently and previously to the publication of M. Le Verrier, and not with the intention of interfering with his just claims to the honors of the discovery, for there is no doubt that his researches were first published to the world, and led to the actual discovery of the planet by Dr. Galle, so that the facts stated above cannot detract, in the slightest degree, from the credit due to M. Le Verrier.[31]

At last Adams received some measure of acclaim. Challis and Airy were sharply criticized. Why hadn't Challis begun a search in September 1845 when Adams first gave him the planet's position? Challis' answer was lame—he lacked confidence in theoretical calculations. Why hadn't Challis reviewed his search data? "[P]artly," he said, "because I thought the probability of discovery was small till a much larger portion of the heavens was scrutinised, but chiefly because I was making a grand effort to reduce the vast number of comet observations which I have accumulated; and this occupied the whole of my time when I was not engaged in observing." No one doubted that Challis had told the truth, but the truth cost him his credibility as a scientist.

Airy stood convicted even in the minds of his friends of "unreasonable incredulity and apathy toward Adams," as his colleague Adam Sedgwick, a geologist, informed him:

> Had the results communicated to you and Challis been sent to Berlin, I am told, they came *so near the mark* that to a certainty the new planet would have been made out in a very few weeks, perhaps a very few days, and the whole business settled in 1845—Adams the sole, unadvised, unassisted discoverer. . . . To say the very least of it, a grand occasion has been thrown away.[32]

And that was from a friend writing to him. Others were far less restrained and less private in their judgment of Airy.

But Airy's high opinion of himself sustained him through the crisis. As always, he felt he had done his job with distinction.

Recognition

It was now annual awards time, and the Royal Society gave its Copley Medal for 1846 to Urbain Jean Joseph Le Verrier for the discovery of Neptune. It had made joint awards before but now declined to honor Adams because it considered his claims too uncertain for notice. The Royal Astronomical Society, having just completed its extraordinary investigation, decided in exhaustion not to award its Gold Medal to anyone in 1846.

Gradually, through the early months of 1847, more of the facts emerged. The hurt feelings and wounded national pride slowly healed. Newspapers called off their vendettas. The accomplishments of Adams and his gentlemanly behavior through the entire proceeding began to be appreciated.

In June 1847, at a meeting of the British Association for the Advancement of Science, Adams and Le Verrier met one another for the first time at a party given by John Herschel. Some guests held their breath. Within minutes, Adams and Le Verrier were off by themselves talking animatedly and laughing. They formed a lifelong friendship.

In 1848, trying to make up for its injustice two years earlier, the Royal Society gave Adams its Copley Medal. The Royal Astronomical Society, too, was ready for a reassessment, but its response was peculiar. For 1848, it gave no Gold Medal but instead awarded "testimonials" to 12 scientists, including Adams, Le Verrier, and Airy, while omitting Galle, d'Arrest, and Challis.

History has been generally kind to Adams. Textbooks and reference works today usually credit Adams and Le Verrier or Adams, Le Verrier, and Galle with the discovery of Neptune, although d'Arrest is still almost always overlooked.

The discovery of Neptune by mathematical calculation—the planet found on a sheet of paper—was the greatest triumph for gravitational theory since 1758, when the comet of 1682 returned just as Edmond Halley had predicted in 1705 using Newton's laws. Ever since, that object has been known as Halley's Comet. Other historians of science argue that the discovery of Neptune was the greatest triumph for Newton's gravitational theory since its publication in 1687.

The discovery of Neptune in 1846 was a turning point in the history of astronomy, as mathematics and theory now took the lead in the way astronomical research was pursued. The greatest epoch, begun with William Herschel's discovery of Uranus in 1781, of discoveries and progress in astronomy led by amateur astronomers was ending. Amateurs would still make contributions, but astronomy was more and more a highly specialized mathematical and physical discipline.

And what did life bring for the major participants in one of the most dramatic and scientifically significant events in the history of astronomy?

Johann Gottfried Galle continued in the position of special assistant that Encke had created for him at the Berlin Observatory in 1835. In 1851 he accepted appointment as professor of astronomy and director of the observatory at the University of Breslau. It was wrenching to leave Encke and the well-equipped Berlin Observatory for a modestly equipped observatory in a provincial town. But there Galle remained for the rest of his life, contributing significantly to binary star, comet, and meteor studies, and when the city lights and obsolete equipment rendered his observatory useless, he shifted with effectiveness into the study of meteorology. When he died in 1910 at the age of 98, he was still sharp of mind and active in astronomy, revered by three generations of Breslau students, and had the unique distincition of having observed Halley's Comet as a professional astronomer at two different apparitions—in 1835 and 1910.

Heinrich d'Arrest was not mentioned in Encke's report on the discovery of Neptune at the Berlin Observatory in 1846. Following his internship at the Berlin Observatory, d'Arrest moved on to the Leipzig Observatory in 1848, where he continued his fine work on comets and asteroids. In 1858 he was named professor of astronomy and director of the new observatory at the University of Copenhagen. There he expanded his scientific reputation with his studies of nebulae and his exploration of the new field of spectroscopy until his untimely death at the age of 52. It was not until 1877, two years after his death, that Galle finally clarified the important role that d'Arrest had played in the detection of Neptune.

George Biddell Airy, astronomer royal, stubbornly rode out the criticism of his role in the Neptune scandal. There is no indication that he ever felt ashamed or regretful or even uncertain of his treatment of Adams or his handling of the quest for Neptune. Airy was impervious because he never doubted himself. Under his direction the Royal Observatory continued to be a model of efficiency as a source of accurate astronomical

Thereafter

Johann Galle in old age
Archive of the Archenhold Observatory

George Airy in 1852, six years after Neptune was discovered
National Maritime Museum Greenwich

tables for British navigators. He remained unpopular with his assistants because he treated them as drudges and tolerated no independent thought. He trained no young astronomers. He was an organizer and a bureaucrat rather than a scientist. In 1881, at the age of 80, he retired from duties as astronomer royal and died, still organizing his papers, in 1892 at the age of 90.

It had been a swift and promising ascent for James Challis from senior wrangler and First Smith's prizeman as an undergraduate at Cambridge to Plumian Professor of Astronomy and director of the Cambridge Observatory in 1836 at the age of 32. Yet no notable science appeared. The Neptune incident focused the attention of the faculty at Cambridge and astronomers in England and abroad to the fact that Challis was something of a kook. His ideas and scientific claims became ever more overblown, perhaps in an effort to extricate himself from the Neptune failure. He claimed that he had generalized Newton's law of universal gravitation to include all physical forces and was about to publish. He never did. Only his friendliness and Cambridge tradition dissuaded critics from stripping him of his distinguished position as Plumian Professor of Astronomy. He did surrender his directorship of the Cambridge Observatory to Adams in 1861. As one historian observed: "Challis was a spectacular failure as a scientist, and ironically, this failure has immortalized him."[33] He died in 1882, just before his seventy-ninth birthday.

Honors continued to pour in for Le Verrier. Professorial chairs in celestial mechanics and astronomy were created for him at the Sorbonne. He continued his study of planetary perturbations for the remaining 31 years of his life, refining the orbits and masses of the planets. He completed this vast project of more than 4,000 pages one month before his death. Le Verrier hoped that other planets would be found the way he and Adams found Neptune.

With typical tenacity, Le Verrier also returned to the problem of Mercury's motion that had stymied him in 1843. Mercury's perihelion was advancing along its orbit as a rate of about 9 minutes 26 seconds a *century*—less than one-third the apparent diameter of the Moon as seen from Earth. In 1843, Le Verrier had been able to explain about 90 percent of this motion by the gravitational perturbations on Mercury of the other planets in the solar system. But 38 (now known to be 43) seconds of arc per century defied his efforts at a complete solution based on the gravity of known objects.

In 1859, 19 years after Arago had handed him the problem, Le Verrier offered an answer. The codiscoverer of the outermost known planet proposed that inside the orbit of Mercury lay an undiscovered innermost planet—or a group of asteroids.

Soon after he published his results, Le Verrier received a letter from

Urbain Jean Joseph Le Verrier
Courtesy of Special Collections, San Diego State University Library

a French country doctor and amateur astronomer named Edmond Modeste Lescarbault who claimed to have seen this planet passing across the face of the Sun. Le Verrier rushed to meet Lescarbault and was convinced that an intra-Mercurian planet had been found, duplicating his planet-predicting triumph with Neptune. A name appropriate for a planet so close to the Sun was already waiting—Vulcan, the Roman god of fire.

Once again, it seemed, Le Verrier had saved the integrity of Newton's law of universal gravitation. But the sighting was not confirmed by other observers at the time or duplicated by later observers. Through the years, various amateur and professional astronomers claimed to have seen Vulcan transiting the Sun or near the Sun during a total eclipse, but these reports conflicted with Lescarbault's sighting and with one another, or they were better explained as misidentified stars.

Ironically, in the carefully observed and analyzed motion of Mercury, astronomers this time had uncovered a real defect in Newton's law of gravity. The problem of Mercury's orbit was not solved until Einstein published his General Theory of Relativity in 1916. The excessive precession of Mercury's perihelion was the most detectable difference in solar system motion between the predictions of Newton and Einstein. And the anomaly that Le Verrier had pointed out provided one of the crucial proofs that Einstein was correct.[34]

Following the death of Arago, Le Verrier became director of the Paris Observatory in 1854. Not only did he continue to lead astronomy from that position, but he was one of the founders of modern meteorology. He set up an international network to warn sailors of approaching storms.

Le Verrier was admired for his absolute integrity in scientific matters, but people close to him found him edgy and authoritarian. His tenure at the Paris Observatory was stormy. He was dismissed in 1870 for repressing staff creativity and devoting too little of observatory funds to astronomy. He was reinstated in 1873 when his successor and enemy Charles Eugène Delaunay died. Thereafter Le Verrier devoted full attention to celestial mechanics. He died of a progressive liver ailment in 1877 at the age of 66.

Appropriate recognition for Adams was slower in coming. His fellowship at St. John's College, Cambridge, expired in 1852 because he had not joined the clergy. He became a fellow at Pembroke College, Cambridge, in 1853. He applied for the position of superintendent of the *Nautical Almanac* in 1853 but was not chosen.[35] Finally, in 1858, he was appointed professor of mathematics at the University of St. Andrews. Upon that appointment, Cambridge University repented its snub and, in that same year, brought Adams back as Lowndean Professor of Astronomy and Geometry. In 1861, Challis was eased out as director of the Cambridge Observatory and Adams was appointed in his place.

John Couch Adams in middle age
The Master & Fellows of St. John's College, Cambridge

In 1863, at the age of 44, Adams married Eliza Bruce from Dublin. They had no children. He was the first president of the Association for Promoting the Higher Education of Women in Cambridge and helped in the establishment of Newnham College in 1880, the first residence for women at Cambridge. He was one of the first professors to open his lectures to women.

Following his codiscovery of Neptune, Adams continued to do distinguished work in astronomical mathematics, bringing him additional renown—and controversy. He carefully studied the gradual acceleration and deceleration in the Moon's motion evident in historical records of eclipses and caused by the changing ellipticity of the Earth's orbit. He found that the analysis of Laplace, the great French celestial mechanician, had been faulty. This caused very hard feelings among French astronomers until, by 1861, independent investigations showed that Adams was right.

He was offered knighthood in 1847 and the position of astronomer royal when Airy retired in 1881 but declined both. The modesty that had cost him fame early in his career endeared him to friends and colleagues throughout his life. John Couch Adams died in 1892 at the age of 72.

CHAPTER 5

Percival Lowell and Planet X

*[On possible names for Neptune]: "The name 'Janus'
would imply that this planet is the last one in the solar
system, and there is no reason to believe that this is so."*
U. J. J. Le Verrier
(September 1846)

*"[A]fter thirty or forty years of observing the new planet
[Neptune], we will be able to use it in turn for the
discovery of the one that follows it in order of distance
from the Sun."*
U. J. J. Le Verrier
(c. 1846)

*[On the clustering of certain comet orbits]: "This can
hardly be an accident; . . . it means a planet out there as
yet unseen by man, but certain sometime to be detected
and added to the others."*
Percival Lowell
(1903)

The discovery of Uranus and Neptune tantalized astronomers with the
prospect that still more distant planets lay undetected. Le Verrier believed
that analysis of the motion of Neptune would reveal the existence of
another planet, but not in his lifetime. Neptune crept along its orbit so
slowly that even at the beginning of the twentieth century its path was
not known with sufficient precision to reveal the perturbations of an outlying planet.

The discovery of Neptune solved the mystery of the anomalous motion of Uranus—almost. When the actual motion of Uranus was compared to predictions that took account of all known gravitational disturbances, tiny discrepancies remained—only a few percent the size of the errors that enabled Adams and Le Verrier to calculate the existence and position of Neptune. Were these "residuals" meaningful? Did they reveal the existence of yet another planet, or were they just minute errors in measuring the exact location of Uranus, resulting perhaps from inaccurate positional data for the background stars? The obstacles to calculating and searching for trans-Neptunian planets were so formidable that few astronomers tried. Many doubted that a planet lurked beyond Neptune. Still, an occasional astronomer attempted some crude calculations that suggested the unexplained inconsistencies in the motion of Uranus (the residuals) pointed to a more remote planet.[1]

The Patrician Amateur

Percival Lowell
Lowell Observatory photograph

On March 13, 1855, Percival Lowell was born into a wealthy Boston family, the oldest of five children. His brother Abbott Lawrence Lowell became president of Harvard University. His youngest sister Amy became a famous poet. Percival graduated with honors in mathematics from Harvard and amassed his own fortune during six years of business work for his grandfather's wide-ranging enterprises, which included cotton mills, financial institutions, and utility companies. Then, in 1883, he began a ten-year series of extended visits to the Far East as a travel writer. He served as foreign secretary and general counselor for the first diplomatic mission from Korea to the United States.

During a trip to Japan in 1893, Lowell learned that Giovanni Schiaparelli, the noted Italian astronomer, was going blind and had been forced to terminate his observations of Mars. It was Schiaparelli who in 1877 had first described thin, straight, crisscrossing features on the surface of the red planet. He named them *canali,* meaning "channels." Channels could be natural or artificial features. The English-language press, however, translated *canali* as "canals," implying that they were built by intelligent beings.

The controversy over artificial features on Mars had long fascinated Lowell. He returned to Boston and began to use his wealth to establish an observatory whose major effort would be the investigation of the Martian "canals."

Lowell chose as his site a pine-forested mesa one mile west of Flagstaff in remote Arizona Territory at an elevation of 7,250 feet (2,210 meters), where his observatory could "see rather than be seen."[2] His was the first permanent observatory in the world to be purposely situated away from city lights and in a climate tested for good seeing conditions.[3]

The Lowell Observatory opened on June 1, 1894. Lowell, an excellent public speaker and writer, was soon lecturing about life forms on Mars whose canals gave proof of their intelligence and global unity but also demonstrated that this peaceful civilization was doomed by drought. Most astronomers were skeptical of Lowell's interpretations of the Martian features; some were harshly critical. Lowell hired a small cadre of talented scientists to operate his observatory and gave them sufficient time for their own research. They left the flamboyance, controversy, and public relations to Lowell and concentrated on meticulous work that brought its own recognition to the Lowell Observatory. In 1905, the permanent scientific staff consisted of Vesto M. Slipher and Carl O. Lampland. Slipher's younger brother Earl C. Slipher soon joined the full-time staff.

A Trans-Neptunian Planet

But Lowell's attention was not exclusively focused on Mars. All the planets intrigued him. His interest in the possibility of a planet beyond Neptune stemmed from his college training by Benjamin Peirce, who created a long-standing controversy by claiming that the discovery of Neptune was an accident because although the position predictions of Adams and Le Verrier were correct, all their other predictions about the planet's distance, orbit, brightness, and mass were in error. Peirce was wrong. Adams and Le Verrier had correctly used perturbations in the motion of Uranus to pinpoint Neptune.

Now, with the means to conduct a telescopic search, an enduring interest in mathematics, and knowledge of previous efforts to hypothesize a ninth planet, Lowell was ready to attack this high-profile problem. Lampland felt that Lowell sought the "prestige of mathematically predicting a new planet . . . in order to gain more respectability for his theories about Mars" and to enhance the prestige of his observatory, which, because of Lowell's claims for canals on Mars, had become "virtually an outcast in professional astronomical circles."[4]

From 1902 on, Lowell's lectures and writings give evidence of his belief that a trans-Neptunian planet would eventually be detected.

His first search for a ninth planet began in early 1905 and proceeded sporadically as his time permitted for four years. The quest began as a photographic survey along the mean plane of the solar system; later a dimension of theory and calculation was added in an attempt to narrow the search. But the observational and theoretical components of the search were seldom integrated. Each time Lowell made a new computation, he altered the hypothetical planet's predicted location and would send his colleagues in Flagstaff a new position to be searched. The photography, completed in September 1907, was performed by a succession of three

University of Indiana astronomy students. Lowell examined the resulting 440 plates but found nothing. Lowell was worried that his project would be stolen by others, so, unlike his other well-publicized ventures, he kept his search for a trans-Neptunian planet as quiet and secretive as he could, referring to it in correspondence with his staff as the "invariable plane" work because the search was conducted along the mean plane of the solar system, thought to be the most likely orbital path for the suspected planet. Eventually, when he found that the complexities of the search were so great that no one could jump in opportunistically, he dropped his secrecy. By late 1908 he was referring to his trans-Neptunian quarry as Planet X.

Competition

In mid November 1908, Lowell received some startling and troubling news. He attended a lecture by William H. Pickering of Harvard, who for a few months had helped him initiate the Mars observation project when the Lowell Observatory opened in 1894 and who was now a critic of his Mars theories. Pickering had used a graphical plot of the residuals of Uranus to predict the existence and position of a trans-Neptunian "Planet O." This graphic approach had first been used and its value demonstrated by John Herschel to refine the orbit of Neptune and to show how its existence and location had been deduced from the residuals of Uranus. His method first appeared in his popular textbook *Outlines of Astronomy* in 1849.

Pickering's 1908 prediction for Planet O gave it a distance of 51.9 astronomical units, a period of 373.5 years, and a mass twice that of Earth. Its disk was estimated to be about 0.8 arc second in diameter and its magnitude either 11.5 (if reflective like Neptune) or 13 (if less reflective like Mars). Two brief telescopic searches had failed to find the planet.

Pickering sought Lowell's help with the search, but Lowell declined, neglecting to mention that he had been calculating and searching for a ninth planet for more than three years himself. Lowell realized that he had serious competition in his quest—and from a scientific adversary whose calculations seemed to be ahead of his own. The shock of discovering Pickering's interest in a trans-Neptunian planet caused Lowell to plunge ahead with all the energy and resources he could spare from his Mars work.

In March 1905, Lowell had hired a moonlighting U.S. Naval Observatory mathematician named William T. Carrigan to assist him with the extraction and analysis of the residuals of Uranus and Neptune. Lowell initially urged Carrigan on, but in May 1909 dismissed him, saying that he had reached his own theoretical solution to the orbit and location of a ninth planet without using any of Carrigan's four years of calculations. Lowell complained that Carrigan was overly meticulous. The planet, Lowell wrote to Carrigan, lay 47.5 astronomical units from the Sun, had an orbital period of 327 years, a magnitude of 13 or fainter, and a mass two-fifths of Neptune's.

Lowell never published this prediction, nor did he use it to search for a planet. His anxiety about Pickering's lead subsided when he read Pickering's published work and found his approach superficial. In annotating his copy of Pickering's paper, Lowell derided his former colleague's efforts, concluding: "This ninth planet is very properly designated O [and] is nothing at all."[5]

A year was to pass before Lowell returned to the problem of Planet X. He began by assessing the shortcomings of his first search. First, there was a problem of instrumentation. The observatory's 5-inch (13-centimeter) telescopic camera, the last and most successful of four telescopes employed in the search, was inadequate to the task. It recorded a sharply defined field of only about 5 degrees. What Lowell could not know was that the undetected planet was then too far south of the ecliptic to be within the search zone and had a brightness of only 16th magnitude (10,000 times too faint for the unaided human eye to see), at the recording limit of his photographic plates.[6]

Assessment

A second problem with the initial search for Planet X was technique. Lowell's method of examining the plates was unlikely to yield results. The search produced plates of the same region of the sky taken a few days apart so that a planet would reveal itself by its change of position. Lowell would take these glass plates, superimpose them, offset them slightly, and then examine each pair of images with a magnifying glass in search of motion. It was cumbersome, unsystematic, and imprecise.

About a year into the first search, Lampland had suggested that Lowell buy a new invention called a blink microscope for the inspection of the plates. With a blink microscope (also called a blink comparator), two plates could be placed side by side and viewed through an eyepiece that optically superimposed the two plates and magnified them. A shutter inside the system then alternated views so rapidly that the shift from one plate to the other was almost undetectable to the eye. If the plates to be compared were identical in images and exposures and if the plates were aligned so as to superimpose accurately, the blink microscope would show what appeared to be just one plate with nothing moving, jumping, or pulsating. But if an object had moved or changed in brightness from one plate to the next, the object that changed would appear to jump

Percival Lowell observing Venus in the daytime with the 24-inch Clark refractor at the Lowell Observatory
Lowell Observatory photograph

back and forth or pulsate as the shutter inside the blink microscope alternated views, while the rest of the images remained unchanged. Lowell decided against an investment in a blink microscope.

Instead Lowell decided that if he were to have any hope of succeeding, the quest would require the most thoroughgoing mathematical analysis possible. A graphical plot of the residuals of Uranus would not do. He would have to repeat and even amplify the rigor that Adams and Le Verrier had applied in their pursuit of Neptune. He faced, he felt, a problem even greater than Adams and Le Verrier had confronted because for them Neptune was the solution, while for him it was part of the problem.

Lowell was ignoring the fact that Adams and Le Verrier had to invent their own mathematical analysis, whereas he could study their methods and employ one of their already proven systems. Nevertheless, he faced some very real problems in his search for Planet X. He ticked off the difficulties:

(1) The residual errors in the predicted versus actual positions of Uranus were now, with Neptune's influence accounted for, very small.

(2) Neptune had not been observed long enough along its huge orbit for perturbations of its motion by an outlying planet to be evident. Therefore he would have to rely on the presumably smaller perturbations exhibited by Uranus.

(3) A ninth planet might have a highly eccentric orbit, and that orbit might be sharply inclined to the ecliptic. (Here Lowell was using the analogy of Jupiter, Saturn, and their satellites as miniature solar systems. The outer moons of Jupiter and Saturn have markedly elliptical orbits that are notably inclined to the equators of their planets.)

(4) The orbit of Uranus (and any planet) cannot be known with absolute precision because of the constantly changing perturbations on it and the difficulty of astronomical measurements. The resulting inaccuracies are incorporated into any calculation using that planet's position.[7]

Lowell reviewed the differences in the methods by which Adams and Le Verrier had attacked the Neptune problem and then chose Le Verrier's. Adams' approach was "direct and masterful," he thought, but Le Verrier's was "simpler and more complete."[8]

A New Attack

The second search began in July 1910 with a mathematical assault by Lowell, assisted by Elizabeth Langdon Williams, who had been editing publications in his Boston office for at least five years previously. Lowell assumed that Planet X traveled around the Sun in the same plane as Uranus and was 47.5 astronomical units from the Sun, the same distance he had used for his initial search.[9] Using the residuals of Uranus, he sought to calculate the orbital eccentricity, the longitude of perihelion, and the mass of Planet X.

On March 13, 1911, Lowell celebrated his fifty-sixth birthday (and the one hundred thirtieth anniversary of the discovery of Uranus) by sending a telegram to Lampland, his assistant director at the Lowell Observatory, requesting him to begin a new photographic survey along the ecliptic for Planet X. Lowell promised that his computations to focus the search would soon follow. Lampland proposed again that the analysis of the photographs would greatly increase in efficiency and precision if the Lowell Observatory had an instrument like a blink comparator. This time, Lowell bought one immediately and it was shipped back and forth between Boston and Flagstaff as needed.[10]

Lowell's first new computations for the position of the ninth planet were not ready until late April. The photographic search proceeded sporadically because Mars was nearing opposition (the point at which the Earth and Mars are closest that occurs every two years) and Lowell and his colleagues were making their usual plans for detailed studies of the red planet.

In 1911, Pickering published his prediction of the existence of three more trans-Neptunian planets beyond his Planet O, designating them P,

WILLIAM H. PICKERING AND PLANETS O, P, Q, R, S, T, AND U

William H. Pickering (1858–1938) made and published more predictions for the existence of trans-Neptunian planets than any other astronomer.

Like Percival Lowell, Pickering was born in Boston, came from a family of social and intellectual distinction, was well educated (M.I.T.), and had travel and adventure in his blood (Pickering especially liked primitive areas and mountain climbing). Like Lowell, Pickering was a bundle of energy. Besides leading four solar eclipse expeditions as a staff member of the Harvard College Observatory, he set up stations of the observatory in Arequipa, Peru (1891), and Mandeville, Jamaica (1900), and helped in 1894 to build the observatory in Flagstaff for Percival Lowell.

He had more than average scientific accomplishments. He was a pioneer in new celestial photography technology. In 1899 he discovered Phoebe, the ninth moon of Saturn and the first satellite to be found photographically. He also produced the first complete photographic atlas of the Moon (1903).

Unlike his older brother Edward Charles Pickering, director of the Harvard College Observatory, William was less rigorous in his work and prone to sudden enthusiasms. This caused considerable friction between the brothers, and William was never promoted beyond assistant professor.

Approaching age 50, Pickering became more and more consumed with predicting the existence and location of trans-Neptunian planets. Although his 1919 prediction of the position for Planet O was as close to where Pluto was found as was Lowell's prediction for Planet X, Pickering's credibility was not high, because of the number of new planets he

William H. Pickering
Harvard University Archives

predicted, the less-than-rigorous method he had for predicting them, and the speed with which he abandoned or substantially modified his predictions.

In 1924 he retired from the Harvard College Observatory, and the observatory he had established for Harvard in Mandeville, Jamaica, became his private observatory.[1]

---

[1]See especially William Graves Hoyt, "W. H. Pickering's Planetary Predictions and the Discovery of Pluto," *Isis* 67 (1976): 551–654.

Q, and R. Planet Q had a mass 20,000 times greater than Earth, which would have made it 63 times more massive than Jupiter and about one-sixteenth (6 percent) the mass of the Sun—close to a star of minimal mass rather than a high-mass planet. Q, said Pickering, had a highly elliptical polar orbit.

Now, as in 1908, competition with Pickering altered none of Lowell's opinions but caused him to intensify his project. He gave orders to accelerate the photographic search and hired four more mathematical assistants to help him and his chief computer, Miss Williams, in Boston. (In those days, computers were human.) All five assistants were at work in November 1912, but Lowell wasn't. He had been pressing so hard on the problem that late in October he collapsed from nervous exhaustion."[11] It was two exasperating months for Lowell before he could return to work for even short periods.

Waiting for a Telegram

In Flagstaff the photographic search proceeded by examining the areas specified by the continually modified calculations. From 1911 to 1914, Lowell conducted the hunt on his observatory's largest instrument, a 40-inch (102-centimeter) reflector, installed in 1909, but its field of view—only one square degree—was too small for an efficient search. In April 1914, Lowell and his colleagues turned to a 9-inch (23-centimeter) Brashear photographic refractor that they borrowed from Swarthmore College's Sproul Observatory near Philadelphia without initially revealing to Swarthmore the exact nature of the project.

The project's intensive computational phase ended in April 1914. Approximately the same orbits and positions for Planet X were recurring in repeated calculations. Lowell sent two of his mathematical assistants to Flagstaff to step up the photographic search, discharged the others, and retained Miss Williams for whatever calculations might yet be needed. Lowell was optimistic when he and his wife set sail in May for their usual spring vacation in Europe. He telegraphed his observatory a new most probable location to search for Planet X, requesting, "Don't hesitate to startle me with a telegram—FOUND!"[12]

But no such telegram arrived. Many years of calculations and searching had not produced the expected planet. Lowell began to get discouraged. He wrote an account of his theoretical and observational efforts for presentation in 1915 to a meeting of the American Academy of Arts and Sciences. Two weeks before the speech, he notified his staff of a newly revised position for Planet X, adding: "I am going to give my work before the Academy on January 13. It would be thoughtful of you to announce the actual discovery at the same time."

Publication and Response

Percival Lowell at his observatory in Flagstaff
Lowell Observatory photograph

Lowell's paper—logical, mathematical, and conservatively phrased—gathered little attention from the press, the general public, or other astronomers. The academy even declined to publish it, and Lowell was irked at having to issue it as a Lowell Observatory publication at his own expense. With the failure of his search and the lack of interest exhibited in his research on a trans-Neptunian planet, Lowell showed his discouragement in his typical way: He turned to other matters and seldom mentioned or took an active part in the project again. "No news from X?" he inquired in July. An entry in the observation log for October 8, 1915, mentions that "Dr. Lowell paid a visit to the 9-inch." But after Lowell published his efforts to find a trans-Neptunian planet in September 1915, Planet X disappeared from his writings and his interviews with the press.

On July 2, 1916, the entry in the observation log read "Lunch." Sky conditions prevented any further work that night. And that was the end of the second photographic search for Lowell's trans-Neptunian planet. The 9-inch telescope was needed back at Swarthmore.

On the nearly 1,000 plates exposed in this second search were 515 asteroids, 700 variable stars—and 2 images of the ninth planet. On March 19 and April 7, 1915, Thomas B. Gill, one of Lowell's former mathematical assistants in Boston, now an assistant observer at Flagstaff, had recorded the sought-after trans-Neptunian planet as part of the ongoing photographic search, but the object went unrecognized. Lowell had predicted that Planet X would be about magnitude 13; these images were between 15th and 16th magnitude—about five times fainter. They were just barely visible on the plates. Lowell probably never examined these plates himself. At this stage in his career, he left such tedious work to his assistants.

Lowell concluded from this second search that the object he was seeking was fainter than 13th magnitude and that therefore the 9-inch telescope used for the search was probably too small. He also concluded that the distance of the trans-Neptunian planet was less than the 47.5 astronomical units he had previously clung to.[13]

He would never know how close he came to Planet X.

On November 12, 1916, Percival Lowell, 61 years old, died of a massive stroke at the Lowell Observatory.

CHAPTER 6

The Discovery of Pluto

"Lunch."
Final entry in the observation log for the 1905–1916 Lowell Observatory
telescopic search for a ninth planet

"We think he is going to develop into a useful man. He has several good qualities that are going to make up for his meager training."
Vesto M. Slipher on newly hired Clyde Tombaugh
(1929)

"Young man, I am afraid you are wasting your time. If there were any more planets to be found, they would have been found long before this."
Visiting astronomer to Clyde Tombaugh
(June 1929)

In his will Lowell provided quite handsomely for his wife, appointed Vesto M. Slipher as observatory director, and left an endowment of more than a million dollars for the support of his observatory. His staff, despite their less flamboyant scientific styles, respected Lowell and wanted to proceed with his projects, including the search for Planet X. But work was slowed not only by World War I but by the lengthy court battle that ensued when Lowell's widow bitterly contested the will. By the time the litigation ended, legal fees had swallowed a substantial fraction of the estate.

If a new search for Planet X were to be mounted, a larger telescope specifically designed for sky survey work would be needed. As soon as the estate was settled in 1927, the Lowell Observatory arranged for the fabrication of a 13-inch (33-centimeter) photographic refractor, made possible by a gift of $10,000 from Abbott Lawrence Lowell, president of Harvard and Percival's brother. The new telescope was assembled and tested at Flagstaff in 1929 and found to be an exceptionally fine instrument, with its wide field providing extremely clear images across a 12- by 14-degree region.

A Kansas Farm Boy

But there was no one to conduct the search. The Lowell Observatory core scientific staff of Vesto M. Slipher, Carl O. Lampland, and Earl C. Slipher professed confidence in Lowell's computations for a ninth planet, but none had room among his existing projects for the time and tedium entailed in a comprehensive photographic search for Planet X.

Vesto Slipher wanted to hire a new assistant observer who was not a research astronomer, lest the observer's own interests distract him from the trans-Neptunian planet search. In 1928, Slipher had received a letter from a young high school graduate in Kansas named Clyde W. Tombaugh. His summers were taken up by farming with his family, but his winters and all clear evenings were devoted to his hobby of astronomy. He had built three reflecting telescopes with mirrors ranging from 7 to 9 inches (18 to 23 centimeters) in diameter, the last two of excellent quality. His drawings and notes about his observations indicated commitment and precision. Despite Tombaugh's lack of formal astronomical training, Slipher decided to give him a chance and invited him to join the staff as an assistant observer on a trial basis.

Clyde Tombaugh, age 22, arrived in Flagstaff on January 15, 1929, without knowledge of the project he would be assigned. He quickly impressed the staff astronomers by his enthusiasm, his willingness to learn, and his painstaking work. The third search for Planet X began with Clyde Tombaugh at the 13-inch refractor on April 6, 1929.

Tombaugh quickly found that the staff astronomers were so busy that he could not rely on them for improvements to the prescribed search procedure. So, after a few weeks, he took the project ever more in hand and made the needed modifications himself.

Clyde Tombaugh with the 13-inch telescope that he used to discover Pluto. (He is looking through the eyepiece for the guide scope.) This photograph was made in 1931, the year after the discovery.
Lowell Observatory photograph

Planet X was expected to be magnitude 12 to 13 or, as Lowell suspected at the conclusion of his second search, perhaps somewhat fainter. The 13-inch telescope could reach stars of 17th magnitude with a one-hour exposure. But, Tombaugh realized, if seeing conditions were poor, the exposure time would require another 15 to 20 minutes. In order to use the blink microscope to compare star fields in search of an object that shifted position from one picture to the next, it was crucial that the plates be made within about a week of one another and that the exposures be made and developed under as nearly similar conditions as possible. This

The Lowell Observatory administration building as it was when Clyde Tombaugh arrived in 1929
Lowell Observatory photograph

Dome for the 13-inch telescope with which Clyde Tombaugh discovered Pluto
Lowell Observatory photograph

The 13-inch photographic survey telescope that Clyde Tombaugh used to discover Pluto. This astrograph, made possible by a gift from A. Lawrence Lowell, Percival's brother, was being installed at the Lowell Observatory in 1929 as Tombaugh arrived to take up observing duties on a trial basis.
Clyde W. Tombaugh

procedure would assure that the background stars on each plate would be as close to identical as possible during blinking so that a real planet (or asteroid or variable star) would stand out. Tombaugh chose a relatively bright star near the center of each star field to serve as a guide star so that the telescope would remain correctly targeted while it turned to track the westward-wheeling stars as the Earth rotated. His plan was

CLYDE TOMBAUGH GOES TO FLAGSTAFF

Clyde Tombaugh, the eldest of six children, was born on a farm near Streator, Illinois, on February 4, 1906. He enjoyed exploring the heavens with his Uncle Lee and his 3-inch telescope. Clyde virtually memorized the popular astronomy book his uncle loaned him. In 1920 his father and Uncle Lee bought a new 2.25-inch scope from the Sears-Roebuck catalog for them to share.

When Clyde was 16, his family moved to a rented farm near Burdett, Kansas, and his uncle insisted that he take the new telescope with him. So much work needed to be done on the farm that Clyde dropped out of school for a year to help. In 1925 he graduated from Burdett High with a longing to be a college professor. But even a college education seemed out of reach. He was needed for farmwork.

Yet the lure of astronomy was strong. In 1926, Clyde fashioned an 8-inch reflecting telescope. He made the mirror out of ship porthole glass, the tube out of pine boards, and the mount out of discarded farm machinery. But the curvature of the mirror was not very good, and he was disappointed with his view of the features on Mars.

With his father's help, Clyde built a storage and storm cellar that could also provide the stable air needed for telescope-mirror testing. He then made a fine 7-inch reflector and sent it to his Uncle Lee. His uncle paid him, and Clyde plunged the money into a 9-inch mirror of his own. His days belonged to farmwork, but his nights were devoted to observing the skies and carefully sketching the planets. He completed his excellent new telescope in time to enjoy the 1928 close passage of the Earth by Mars.

The growing season of 1928 was developing in-

to one of the best the family had ever known when a sudden hailstorm ruined the crop just before harvest. Clyde decided against a career in farming. He was 22 years old. He needed to get a job to earn some money to help his family through the crisis. He weighed joining the railroad as an apprentice fireman or trying to start a telescope-making business. He sent a few of his meticulous sketches of Jupiter, Saturn, and Mars off to the Lowell Observatory for appraisal. "It was the only planetary observatory I knew of," he confessed years later.[1]

His letter arrived at an opportune time. Vesto Slipher, the observatory director, was looking for a committed young observer to resume the photographic search for a trans-Neptunian planet suspected by observatory founder Percival Lowell and upon which he had invested so much effort. Slipher, his brother Earl, and Carl Lampland, the three professional astronomers on the Lowell Observatory staff, had all grown up doing farmwork. Slipher invited Tombaugh out to Arizona on a three-month trial basis. Clyde's family drove him 30 miles to the nearest railroad station. His father advised him, "Clyde, make yourself useful, and beware of easy women." The Kansas he left had for its state motto *Ad astra per aspera*—"To the stars through difficulties."

Tombaugh arrived in Flagstaff on January 15, 1929, after a train ride of 28 hours. He didn't have enough money to sleep in a Pullman berth. He didn't have enough money for a train ride home.

---

[1]This and the following quotation are from Clyde W. Tombaugh and Patrick Moore, *Out of the Darkness: The Planet Pluto* (Harrisburg, Pennsylvania: Stackpole Books, 1980), p. 25.

to photograph each region three times. The two most nearly identical plates would be blinked against one another. The third plate and pictures taken by a 5-inch (13-centimeter) telescopic camera mounted piggyback on the telescope tube would be kept for comparison and—hopefully—verification.

Tombaugh started in Cancer, to the east of Lowell's predicted location

Clyde Tombaugh with his homemade 9-inch telescope on his family's farm in Kansas. The year is 1928, and Tombaugh is 22 years old.
Clyde W. Tombaugh

Slipher met him at the station, and Tombaugh moved into a bedroom at the observatory.

The lens for the 13-inch telescope to be used for the search had not yet arrived, so Clyde was pressed into service showing tour groups around the observatory, stoking the furnace in the ad-ministration building (as all the staff did), carefully pushing snow off the canvas dome for the 42-inch telescope, and painting the 13-inch telescope tube red.

The lens arrived in February. Slipher coached him through the photographic process, and quickly Tombaugh was on his own. He faced a series of problems that threatened the very precise work necessary. When the clock drive of the telescope (to keep it pointed at the proper star field as the Earth rotates) turned through a particular position, one of the telescope axes slipped slightly, creating double images of each star. Such a photograph was useless for blinking. A second problem was that the glass photographic plates he was using shattered in the numbing cold—and with a crack so loud Tombaugh feared that the expensive telescope lens had broken. It was up to him to find the cause and invent a fix for each problem. He did. For the axis slippage problem, he ran the telescope ahead and then backed it up through the problem point so that there would be no slippage when it ran forward. For the problem of the photographic plates that cracked, he found a new way to fasten them onto the telescope so that the corners could expand or contract before being tightened down.

Slipher was pleased with Tombaugh's progress. "We think he is going to develop into a useful man. He has several good qualities that are going to make up for his meager training. . . . He has a good attitude: careful with apparatus, willing to do anything to make himself useful and is enthusiastic about learning and wants to do observing."[2]

---

[2]Letter of February 7, 1929, to Roger Lowell Putnam, Percival's nephew and sole trustee of the observatory. Quoted by William Graves Hoyt, *Planets X and Pluto* (Tucson, University of Arizona Press, 1980), p. 181.

for Planet X. Each exposure captured 50,000 stars. Then Tombaugh pushed on into Gemini, closer to the Milky Way. Now each plate provided 400,000 stars to be inspected.

Five days into his search, on April 11, 1929, Tombaugh exposed plate number 10 and captured the object he was seeking. He captured it again on April 30, its companion plate for comparison. But the ninth planet went unnoticed by the Lowell Observatory astronomers who were responsible for the blink comparator examination. Plate 10 had cracked in the brutal cold of winter at high altitude. It was still blinkable, but the plates were not sufficiently well matched in background star brightness for successful comparison. Gemini was also too close to the horizon, so that light from that broad region passed through significantly varying amounts of the Earth's atmosphere, dimming the stars closer to the horizon more. Further, Planet X was expected to be significantly brighter than it was. There was little point in spending much time with such unpromising plates.

Blinking

Originally, Tombaugh was supposed to do only the photography. The blink microscope work would be left to the more experienced staff. But by the beginning of summer, 1929, almost no plates had been blinked. The experienced staff had been too busy with other duties. The summer rains set in, reducing Tombaugh's observing time, so Slipher asked him to take over responsibility for plate comparisons on the blink microscope as well. "I was overwhelmed," Tombaugh recalled. "It had become evident to me that the one doing the blinking carried the heavy responsibility of finding, or not finding, the planet."[1]

Clyde Tombaugh using the blink comparator with which he discovered Pluto. This photograph was taken in 1938, as Tombaugh continued his full-sky survey.
Lowell Observatory photograph

Tombaugh spent half of each month at the telescope and the other half, when the presence of even the crescent Moon above the horizon made the sky too bright for such sensitive telescopic photography, at the blink microscope. His sessions at the blink microscope would last three to six hours; beyond that the numbing routine was so exhausting that he could not stay alert. The average star count per photograph was 160,000.[2] Areas of the sky toward the center of the Milky Way yielded plates filled with a million stars. "I came to dread the Milky Way regions," said Tombaugh.

He divided the plates so that he could blink a few hundred stars at a time, examining each one for a telltale change in position. Tiny irregularities in the concentration of silver grains in the photographic emulsions gave rise to dozens of false planet suspects on every plate. There were numerous variable stars that were visible at maximum brightness but invisible at minimum. They too could look like stars that had moved.

But "even more appalling" were the problems caused by asteroids.[3] These minor planets in orbit around the Sun had a motion similar to but generally faster than the planet he was seeking. Yet when the Earth caught up with and passed each minor planet, it would appear to stop, go backward (retrograde motion), and then move forward again. When traveling slowly near a transition point to and from retrograde motion, the asteroids closely mimicked the movement of a trans-Neptunian planet. Often in the course of examining each star on every plate, Tombaugh saw a starlike object that changed position, but each time it proved to be an asteroid.

After much thought, Tombaugh solved the problem by photographing the sky 180 degrees from the Sun's position (part way up the southern sky at midnight). Here the retrograde motion of a distant planet and an

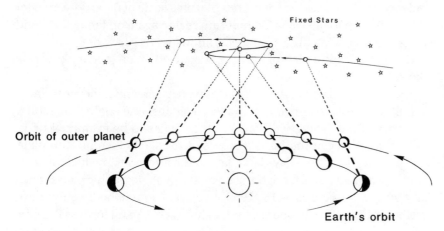

Fixed Stars

Orbit of outer planet

Earth's orbit

As the Earth catches up with and passes a slower-moving outer planet, that planet appears to stop and back up (move westward) in the sky. As the Earth moves ahead of the other planet, it appears to stop its retrograde motion and resume its direct motion eastward among the stars.

asteroid would be most noticeable In fact, at opposition all the asteroids moved enough during the one-hour exposures to create a trail on the photographic plates, thus allowing him to discriminate them from the planet he sought.

Each photographic plate required a one-hour exposure. Each pair of plates required three days to a week or more of blinking. It was maddening work.

But potentially the worst discouragement of all that Tombaugh faced was the skepticism about a trans-Neptunian planet that he heard from visiting astronomers. Most of Lowell's ideas were branded as fanciful, and Pickering's planet predictions were thought to be wild speculations. Tombaugh remembered poignantly when a visiting astronomer privately confided to him, "Young man, I am afraid you are wasting your time. If there were any more planets to be found, they would have been found long before this."

Expanding the Search

By early September, Tombaugh decided to search not just the area where Lowell had predicted the planet would be but the entire zodiac. He had little confidence in Lowell's predictions because they were based in part on the idea that the location of a ninth planet could be at least roughly established by the trans-Neptunian aphelia (far points in their orbits) of a family of comets. He resolved to make his search so thorough that "if nothing was found, I would be able to state such a planet does not exist."

Late in January 1930, Tombaugh was photographing the western portion of the constellation Gemini. His search of the zodiac had now come full circle. He had returned to the region that he began with.

On January 21, he exposed Negative 161, centered on the star Delta Geminorum. The sky transparency was good but heavy atmospheric turbulence provided seeing conditions of 0 on a scale of 0 to 10—very poor.

He returned to the Delta Geminorum region again on January 23, with Negative 165. This time seeing conditions were very good.

He made his third plate of the area, Negative 171, on January 29. Again, the seeing was very good.

The January 23 and January 29 plates were taken under the most similar conditions, so he planned to blink them against one another as soon as time permitted. There were many other plates to be examined first.

Tombaugh began the blink analysis of the Delta Geminorum plates on February 15. Three days later, late in the afternoon, he had made his way through one-quarter of the stars on these photographic plates. Then, at 4:00 P.M. on February 18, 1930, two weeks after his twenty-fourth birthday, Clyde Tombaugh saw a star shifting back and forth as the blink microscope cut the view from one optically superimposed plate to the

The discovery of Pluto. Here are small sections of the plates on which Clyde Tombaugh discovered Pluto on February 18, 1930. The photographs were taken on January 23, 1930 (left), and January 29, 1930 (right). Pluto, marked by an arrow, revealed itself by its motion.
Lowell Observatory photograph

other. It was 15th magnitude—fainter than the object expected. "That's it," Tombaugh said to himself.

> A terrific thrill came over me. I switched the shutter back and forth, studying the images. Oh! I had better look at my watch and note the time. This would be a historic discovery. . . .
>
> For the next forty-five minutes or so, I was in the most excited state of mind in my life. I had to check further to be absolutely sure.

They were no defects on the photographic plates. The two dots didn't look like two different variable stars. The change in position between January 23 and January 29 was right for a trans-Neptunian planet in retrograde motion. He checked for the object on the January 21 plate and on the simultaneous photographs made by the 5-inch telescopic camera mounted piggyback on the 13-inch. The object was real.

For 45 minutes he had been the only person on Earth to know of the existence of a ninth planet. At 4:45 P.M. Tombaugh called across the hall to Lampland that he had found a trans-Neptunian planet. Yes, said Lampland, "I heard the clicking of the comparator suddenly stopped, then a long silence." Lampland immediately began studying the images. "Then," said Tombaugh,

> I walked down the hall to V. M. Slipher's office. Trying to control myself, I stepped into his office as nonchalantly as possible. He looked up from his desk work. "Dr. Slipher, I have found your Planet X." I had never come to report a mistaken planet suspect. He rose right up from his chair with an expression on his face of both elation and reservation. I said, "I'll show you the evidence."
>
> He immediately hurried down the hall to the comparator room. I had to step lively to keep up with him.

"[T]he air was tense with excitement," Tombaugh recalled.[4] Vesto Slipher insisted that there should be no announcement until the obser-

Plate 1. Voyager 2 at Nepture and Triton (Artwork by Don Davis)
NASA/Jet Propulsion Laboratory

Plate 2. Astronomer Dale P. Cruikshank proposes that Neptune's moon Triton may have seas of liquid nitrogen. Artist Ken Hodges depicts Neptune looming over this chilly landscape.
NASA/Jet Propulsion Laboratory

Plate 3. Voyager 2 launch, August 20, 1977
NASA/Jet Propulsion Laboratory

Plate 4. Voyager 2 images of Uranus. On the left is Uranus in real color. No atmospheric details are visible. On the right is a computer-enhanced image of Uranus, with false color added to bring out detail. Now the atmosphere shows bands that run parallel to the equator. The south polar atmosphere of Uranus (pointed toward and a little to the left of us) is slightly darker in color.
NASA/Jet Propulsion Laboratory

Plate 5. A *Voyager 2* image of Uranus, with computer enhancement and false color added, shows a very large cloud (more than 5,000 miles [8,000 kilometers] long) in the planet's atmosphere in the one o'clock position. The south pole of Uranus is pointed almost directly at us, so this cloud lies near the equator. (The doughnut-shaped rings in the picture are caused by dust in the camera.)
NASA/Jet Propulsion Laboratory

Plate 6. Uranus and its ring system seen from Miranda. Separate *Voyager 2* images have been used to simulate what a spacecraft near Miranda would see of Uranus.
NASA/Jet Propulsion Laboratory

Plate 7. From above the north pole of Uranus, artist Ron Miller visualizes the narrow Uranian rings which circle the planet's equator.
© *Ron Miller*

Plate 8. Voyager 2 found fault valleys up to 12 miles (20 kilometers) deep on the Uranian moon Miranda. Artist MariLynn Flynn imagines astronauts at the bottom of this canyon with a crescent Uranus in the sky.
© *1987 MariLynn Flynn*

PLATE 1

PLATE 2

PLATE 4

PLATE 5

PLATE 6

PLATE 3

PLATE 7

PLATE 8

PLATE 9

PLATE 10

PLATE 11

PLATE 12

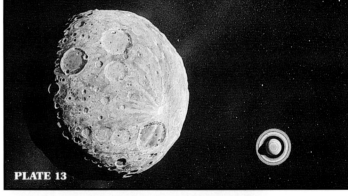

PLATE 13

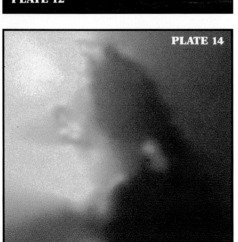

PLATE 14

PLATE 15

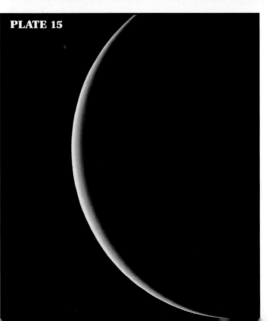

Plate 9. CRAF (Comet Rendezvous Asteroid Flyby) mission near a comet as surface ices vaporize. Astronomer and artist William K. Hartmann imagines that the comet's high carbon content gives it a reddish color.
© *Painting by William K. Hartmann*

Plate 10. The formation of the solar system: Planetesimals accrete to form planets.
© *Painting by William K. Hartmann*

Plates 11 and 12. Charon casts a shadow on Pluto. Here, artist Joe Shabram views the eclipse from two angles: from Charon and at a distance from both. Pluto, with polar ice caps, reflects more light than Charon. Nighttime on Charon is not completely dark because of sunlight reflected from Pluto. That sunlight, in turn, is reflected from Charon back to Pluto and keeps the shadow cast by Charon from being completely black. Note the size of Charon (and its shadow of equal diameter) compared to Pluto.
© *1988 Joe Shabram*

Plate 13. Astronomer and artist William K. Hartmann depicts Chiron, the outermost known asteroid, passing over the pole of Saturn. Chiron roams from just inside the orbit of Saturn to almost the orbit of Uranus.
© *Painting by William K. Hartmann*

Plate 14. The nucleus of a comet was seen for the first time when the European Space Agency's *Giotto* spacecraft used its Halley Multicolor Camera to photograph Halley's Comet during its flyby on March 13, 1986. Comet Halley's nucleus is jet black in color and irregular in shape—about 10 miles (15 kilometers) wide. Gases from vaporizing surface ices carry dust grains outward in jets. Comets may be planetesimals left over from the formation of the outer planets and their moons.
© *1986 Max-Planck-Institut fur Aeronomie, Federal Republic of Germany; courtesy of H. U. Keller*

Plate 15. Looking back at Uranus, *Voyager 2* took this farewell picture as it began its 3.5-year journey to Neptune. Uranus is never visible as a crescent from Earth.
NASA/Jet Propulsion Laboratory

vation could be thoroughly confirmed. "Don't tell anyone until we follow it for a few weeks," said Slipher. "This could be very hot news."

Tombaugh didn't even tell his parents. They found out about their son's discovery after the formal announcement almost a month later, when the editor of their county newspaper called.

Confirmation

**Carl O. Lampland,
assistant director of
Lowell Observatory**
*Lowell Observatory
photograph*

To confirm the existence of the new planet, the Lowell astronomers decided to rephotograph the region where the suspect had been found. But that night, February 18, was overcast. Tombaugh, trying to fill the time, went off to Flagstaff to see Gary Cooper in *The Virginian*. "There would never be another day like that one," Tombaugh reckoned late in life.

The weather cooperated at least to some degree on the next four nights. Tombaugh captured the object with the 13-inch telescope on February 19, 21, and 22. On February 20, Slipher, Lampland, and Tombaugh examined Pluto visually with the 24-inch (61-centimeter) refractor. The next night, Lampland photographed the suspect with the 42-inch (107-centimeter) reflector[5] under seeing conditions listed as poor. The object had no discernible disk, the tip-off of a planet. Yet it lacked the tail, fuzzy coma, and other features of a comet. Sadly, there was no satellite visible, which would have quickly confirmed it as a planet and would have offered a means to measure its mass.

Although everyone on the Lowell Observatory staff was confident about the discovery, Slipher delayed the announcement so that further confirmatory observations could be made. He also wanted to get a head start on the computation of the planet's orbit so that the Lowell Observatory could be the first to publish it. He wanted as well to gather as much information on the new planet as possible before the observatories with larger telescopes overpowered their work. Slipher invited John A. Miller to come to Flagstaff to compute the planet's orbit. Miller, director of the Sproul Observatory at Swarthmore College, was a specialist in celestial mechanics and had taught the Sliphers and Lampland at Indiana University.

On March 12, 1930, at 10:00 P.M. Arizona time—midnight in Boston—Slipher sent a telegram to the Harvard College Observatory, clearinghouse for the announcement of astronomical discoveries. The discovery of a trans-Neptunian planet was announced the next day—March 13, 1930. It was the seventy-fifth anniversary of the birth of Percival Lowell and the one hundred forty-ninth anniversary of William Herschel's discovery of Uranus. The ninth planet had been found within 6 degrees of the spot predicted by Percival Lowell in 1915.

The story was front-page news around the world, and the scale of public interest amazed the Lowell staff. Harlow Shapley, director of the Harvard College Observatory, was scheduled to deliver a talk on astronomy to a small audience in Philadelphia on the evening of March 14. He decided to include mention of the new planet. The lecture had to be moved to a ballroom so that a thousand extra people could be accommodated. When he showed a slide of Percival Lowell, the audience gave a thunderous ovation—greater, Shapley felt, than any Lowell had received in his lifetime.[6]

But Shapley and others were annoyed at and soon critical of the Lowell Observatory for withholding all but one position for the new planet. Slipher had released enough information for other observatories to confirm the planet's existence but not enough so that other astronomers could compute the orbit of Pluto without accumulating their own positional observations over a period of a month or so. By hoarding their January-through-March observations, the Lowell Observatory astronomers had, they hoped, given themselves enough data so that their orbital computation could be better than others.

Vesto M. Slipher, director of the Lowell Observatory after Lowell's death
Lowell Observatory photograph

Even so, two orbital computations appeared before that of the Lowell Observatory. The first (released April 7) was from Armin Otto Leuschner, Ernest C. Bower, and Fred L. Whipple at the University of California, Berkeley, which gave a cometlike parabolic orbital approximation for the new planet. Its distance from Earth, they computed, was 41 astronomical units and the inclination of its orbit was 17 degrees. On April 8 a cir-

LOWELL'S CALCULATIONS AND MY SEARCH
by Professor Clyde W. Tombaugh

It has been stated by some writers of the Pluto story that Percival Lowell's calculations greatly aided my finding Pluto. Quite to the contrary, when I found out in the latter part of 1929 how Lowell had drastically changed his predicted position of Planet X from Libra to Gemini, this indicated to me that considerable uncertainty was involved, and I could not take the prediction seriously.

At the start of my trans-Neptunian planet survey at the Lowell Observatory in the spring of 1929, I photographed the region where Lowell had predicted Planet X to be and the two Sliphers blinked the three pairs of plates I had taken, spanning Gemini. As it turned out, the Pluto images were on the Delta Geminorum pair, but the Sliphers missed them. Since the Sliphers failed to detect the Pluto

images on these plates, I wrote Gemini off and concluded that Planet X might be anywhere in the zodiac belt, if indeed Planet X existed at all. I intended to thoroughly search the entire zodiac. After I became experienced in blinking in the fall of 1929, I realized that the Sliphers had scanned the plates too hastily and that I should rephotograph the Gemini regions again, this time under the proper observational strategy, and then thoroughly blink the new plates.

I rephotographed Gemini when it returned to opposition in January 1930. I was blinking one of the pairs of plates when I found Pluto two-thirds of a degree east of the star Delta Geminorum on February 18, 1930.

cular orbital approximation from the Cracow Observatory in Poland was published in the United States.

On April 12, 1930, Slipher was ready and the Lowell Observatory released its provisional orbit along with a number of planet positions. This work, led by Miller, calculated that the ninth planet had an extremely elliptical orbit (eccentricity of 0.909) and a period of 3,000 years. These initial calculations cast some doubt on whether the object was a planet or not. The problem was that a decent orbit was hard to calculate when the object had moved such a tiny amount along its path. On May 9, British astronomer Andrew C. D. Crommelin reported that the new planet was part of a field photographed at the Royal Observatory of Belgium at Uccle on January 27, 1927, thereby extending the recorded arc of the planet's motion by more than three years. The new calculation showed a much less elliptical orbit (eccentricity 0.287) and a period of 265.3 years. The trans-Neptunian object now had a clearly established planetary nature, and this status was never again seriously challenged. These figures, now close to the actual values for the ellipticity and period of Pluto's orbit, were also close to Lowell's prediction for Planet X.

Naming the Planet

Venetia Burney, age 11, from Oxford, England, was studying mythology in school when she heard about the discovery of the ninth planet and proposed the name Pluto.

Mrs. E. Maxwell Phair

Meanwhile, suggestions for names flooded the observatory. Mrs. Lowell first proposed Zeus, then decided that the old gods were worn out and that the planet should be named Percival. Then she changed her mind again and proposed her own first name—Constance. In a final effort, she urged the name Planet X. Again, Slipher refused to be pressured into a quick announcement. He personally favored the name Minerva, goddess of wisdom, but decided that that name was too firmly attached to an asteroid. Acceptable alternatives, he felt, were Cronus and Pluto. But Cronus was suggested by "a certain detested egocentric astronomer," so this name was discarded.[7] (The rejected name had come from Thomas Jefferson Jackson See, an early and abrasive Lowell Observatory staff member.) The name Cronus (or Kronos) was also considered less than optimal because Cronus is the Greek equivalent of Saturn, who was already well represented in the sky.

The name Pluto was officially proposed by Slipher on May 1, 1930. Pluto had actually been one of the first name suggestions received at the Lowell Observatory. It had come from Venetia Burney, an 11-year-old schoolgirl in Oxford, England, who was learning about Greek and Roman mythology in school and thought that such a dim and gloomy planet should be named for the god of the underworld.[8] As a symbol for Pluto, Slipher proposed PL, formed from the letters P and L, the first two letters in the word Pluto. They were also the initials of Percival Lowell.

William H. Pickering, author of many outer-planet predictions, was initially unhappy with the choice of Pluto as a name. From his retirement home and private observatory on the island of Jamaica, he complained that he had intended to use the name Pluto when his Planet P was found (although he had never published his intention). Eventually he reconciled himself to the name Pluto and its PL symbol, commenting to a visiting astronomer, "That's a good name—Pickering-Lowell!"[9]

Whom to Credit

Once Pluto was shown to have a planetary orbit, the question arose as to whether Lowell or Pickering had been more accurate in his predictions. Granted, it was solely Lowell's calculations that had spurred the search that led to the discovery of Pluto at his observatory, but Pickering's fairly similar predictions had lent some credibility to the search.

Lowell's predicted orbital elements for Pluto were reasonably close and, in fact, except for the critical element of location, better overall than Adams' and Le Verrier's predictions for the orbit of Neptune. Pluto was found within 6 degrees of the locations specified by both Lowell and Pickering (in his 1919 prediction of Planet O). Pickering was actually 0.1 degree closer.

Clyde Tombaugh drew this diagram to compare the actual orbit of Pluto with those predicted by Lowell in 1914 and Pickering in 1928. The orbits are approximated by circles with the positions of their centers noted.
Clyde W. Tombaugh

Lowell's prediction was more nearly correct than Pickering's in the planet's distance from the Sun, orbital eccentricity, longitude of perihelion, perihelion date, and period. Pickering's 1919 prediction was closer to the mark in orbital inclination, longitude of ascending node, mass, and magnitude.

Pickering had revised his Planet O prediction in 1928 so markedly that the anticipated object was very different from his 1919 planet. The 1928 prediction foresaw an object with a mass of only 0.75 (still closer to the truth) and with a most unusual orbit that crossed inside of Neptune's. All other aspects of Pickering's 1928 Planet O prediction were conspicuously wrong.

At first it looked as if the roles of Adams and Le Verrier had been replayed: Two dedicated researchers had independently solved a problem that the scientific community had doubted could be solved, and their insight had been vindicated by a striking discovery.

But as more data flowed in, the question became whether Pluto, found where Lowell calculated it would be, was actually the Planet X he had had predicted. The overwhelming problem, right from the beginning, was one of mass. Pluto showed no disk to observers, even with the largest telescope at that time—the 100-inch (2.5-meter) one on Mount Wilson. Thus Pluto had to be small, far smaller than the 6.6 Earth masses that

LOWELL'S AND PICKERING'S PREDICTIONS FOR PLUTO[1]

Orbital Elements	Lowell's Planet X 1914	Pickering's Planet O 1919	Pluto 1930
Mean distance (astronomical units)	43.0	55.1	39.5
Eccentricity	0.202	0.31	0.248
Inclination	about 10°	about 15°	17.1°
Longitude of node	(not predicted)	about 100°	109.4°
Longitude of perihelion	204.9°	280.1°	223.4°
Period (years)	282	409.1	248
Perihelion date	February 1991	January 2129	September 1989
Longitude (1930.0)	102.7°	102.6°	108.5°
Mean annual motion	1.2411°	0.880°	1.451°
Mass (Earth=1)	6.6	2.0	less than 0.7
Magnitude	12–13	15	15

[1]Based on William Graves Hoyt, *Planets X and Pluto* (Tucson: University of Arizona Press, 1980), p. 221.

Lowell had predicted; even smaller than the 2.0 Earth masses that Pickering (1919) had predicted. The mass could certainly be no greater than that of our planet. Then how was it possible that distant Pluto could have noticeably disturbed the motion of Uranus, a billion miles (1.6 billion kilometers) away at its closest—as far as the Earth is from Saturn?

Yet, could it be that two independent researchers would so nearly coincide in their predictions and that their predictions would so nearly coincide with reality unless their work was valid?

The controversy continued for many years.[10] Throughout those years new telescopes and techniques kept whittling away at the size of Pluto. Measurements by University of Hawaii astronomers in 1976 found spectroscopic evidence of methane frost on its surface. Such an icy covering could give Pluto a high albedo (reflectivity). But a high reflectivity meant that the planet was still smaller than previously expected—smaller than our Moon. Under those circumstances, Pluto's mass could be only a few thousandths that of Earth.

CLYDE TOMBAUGH AFTER PLUTO

Clyde Tombaugh took leave from the Lowell Observatory each school year beginning in the fall of 1932 to formally study astronomy at the University of Kansas on a scholarship. He entered as a 26-year-old freshman and tried to enroll in the freshman astronomy class. But the head of the department was adamant: "For a planet discoverer to enroll in a course of introductory astronomy is unthinkable."[1]

At the end of spring semester 1934, he married Patricia Irene Edson from Kansas City, and they spent their honeymoon in Flagstaff. They had two children. In 1939, Tombaugh received a master's degree in astronomy.

As his planet search through the entire sky from Flagstaff neared completion, Tombaugh was drafted, in July 1943, the middle of World War II, to teach navigation for the Navy at Arizona State College at Flagstaff.

Following the war he was visiting professor of astronomy at UCLA for a year and then, in 1946, went to the White Sands Missile Range in New Mexico as the supervisor of the Optical Tracking Section. There he developed a tracking telescope that could see the tail fins on V2 rockets at an altitude of 100 miles (160 kilometers). Tombaugh's work led to the tracking cameras that follow manned and unmanned rockets from launch to orbit today and that, aboard spacecraft, look down on Earth to provide pictures with incredible detail.

In 1955 he was appointed associate professor at New Mexico State University, where he helped to found the Department of Astronomy. He became a full professor in 1956. He has been professor emeritus since 1973 and continues to inspire students with his lectures on astronomy.

In honor of his service and achievements, New Mexico State University is establishing a Clyde Tombaugh Scholars Endowment to assist postdoctoral astronomy students and an endowed professorship, the Clyde Tombaugh Chair in Astronomy.

---

[1]Dinsmore Alter, quoted in Clyde W. Tombaugh and Patrick Moore, *Out of the Darkness: The Planet Pluto* (Harrisburg, Pennsylvania: Stackpole Books, 1980), p. 160.

A Bulge

The matter was settled in June 1978 when U.S. Naval Observatory astronomer James W. Christy was examining pictures of Pluto taken on April 13 and 20 and May 12, 1978. These photographic plates had been exposed at the Naval Observatory's station in Flagstaff, Arizona, only four miles from the Lowell Observatory, where Pluto was discovered. Christy found a bulge in the image of Pluto—a bulge that had shifted position on the second plate. Christy immediately interpreted this bulge as a satellite in revolution around Pluto, and examination of earlier survey photographs verified it.

As a name for Pluto's moon, Christy chose Charon, the boatman in Greek mythology who ferries the souls of the dead across the rivers of woe and lamentation to Pluto's underworld. Beyond its mythological appropriateness, Christy was especially fond of the name because his wife's name is Charlene—nickname: Char.

The presence of a moon in orbit about Pluto allowed for the calculation of the mass of the combined Pluto/satellite system. This calculation, by Christy's Naval Observatory colleague Robert S. Harrington, showed a mass of about two-tenths of one percent (0.002) of the Earth. Pluto was only about 20 percent the mass of our Moon.

The mass of the Pluto-Charon system was hopelessly inadequate to produce measurable gravitational perturbations on Uranus or Neptune. Pluto could not be Percival Lowell's Planet X. The planet found was not the planet sought. What had seemed to be another triumph of celestial mechanics turned out to be an accident.

Or, rather, Pluto had been found not by theory but by the intelligence and thoroughness of Clyde Tombaugh's search.

CHAPTER 7

Toward Uranus

"Fate often smiles kindly upon great endeavors."
Charles Kohlhase, *Voyager* Mission Design Manager

A flashlight provides more light in one second than Uranus has provided to Earth in the 200 years since it was discovered by Herschel in 1781.[1] No wonder that two centuries passed before Uranus was revealed in detail.

A Long Wait

Watching Uranus as scarcely more than a dot in the skies had told astronomers about its orbit. Watching its moons in orbit had told astronomers the mass of Uranus and that the planet was lying on its side as it revolved about the Sun. Nearly every time a large new telescope went into operation, it was trained on Uranus (and Neptune) to try to see details, to measure a rotation period, and to discover new satellites. Uranus was uncooperative.

Astronomers settled into a routine. The size of Uranus was refined. The orbits and periods of the Uranian satellites were refined. The mass of Uranus, calculated from its satellites' periods, was refined.

Claims to have seen surface markings on Uranus were made and not confirmed and made again later by others. But ever-improving telescopes still could not confirm details on the face of Uranus.

Uranus as seen from Earth is the size of a golf ball seen at a distance of 1.5 miles (2.5 kilometers). With no distinct atmospheric clouds, bands, or storms to provide contrast, the period of rotation for Uranus—the length of its day—was difficult to measure. Even spectroscopy—breaking down

An excellent view of Uranus from Earth, showing the five moons known before the visit of *Voyager 2* in 1986. The satellites are the bright, starlike objects closest to Uranus. Titania is at the one o'clock position; Miranda is at four o'clock, almost lost in the glare of Uranus; Umbriel is at seven; and Oberon and Ariel are at eleven, with Oberon the more distant. The south pole of Uranus is pointed almost directly at us, so these moons revolve like a spot on the hand of a clock. This picture was taken by W. Liller on June 12, 1977, with the 158-inch (4-meter) telescope at the Cerro Tololo Inter-American Observatory in Chile. *National Optical Astronomy Observatories*

light into its component wavelengths to study subtleties such as composition, temperature, and atmospheric flow—was very difficult and prone to error for such a small, faint object at a distance of almost 2 billion miles (3 billion kilometers).

As the bicentennial of the discovery of Uranus neared, the number of moons known for Uranus had risen to five. The atmosphere, as expected from the example of Jupiter and Saturn, was known to be predominantly hydrogen. The pale bluish-green color of the planet could be assigned to a very small percentage of methane in the atmosphere that absorbed red light. Spectroscopy indicated a rotation period for Uranus of about 10 hours 50 minutes.[2]

And that was pretty much all that was known. But the distance and faintness of Uranus did not stop Earth-bound astronomers from using every technique at their disposal to wring more secrets from the tilted world. A big surprise came unexpectedly from a flying telescope.

On March 10, 1977, several independent teams of astronomers were in the Southern Hemisphere to watch as Uranus passed in front of a star. Since the position and speed of Uranus in orbit were well known, a timing of the occultation could provide improved information about the size of the planet. Observation of the way the star's light faded as Uranus eclipsed it could reveal the nature of the planet's atmosphere.

An American team of astronomers, led by James Elliot, was aboard NASA's Kuiper Airborne Observatory, a converted military C-141 Starlifter cargo jet carrying a 36-inch (0.9-meter) telescope and photometry equipment. They were flying over the southern Indian Ocean at 41,000 feet (12,500 meters) to surmount as much as possible of the Earth's turbulent atmosphere and especially its water vapor, which absorbs the infrared light that could provide information not obtainable from the surface of Earth. Because the position of the star to be occulted was not known with the greatest exactitude, the sensing equipment was turned on 47 minutes early to avoid missing the event.

The action began immediately. The light from the star dimmed, brightened, then dimmed again, flickering at least five times. Yet the edge of Uranus had not yet encroached upon the star. And no flaw in the instruments could be found. What was going on? Had a series of undetected Uranian satellites caused the flickers?

As Uranus moved on in its orbit and the star emerged from behind its disk, the flickers repeated themselves in reverse order. Such symmetry could not be the result of moons. The Kuiper Airborne Observatory team

Dimmings

NASA's Kuiper Airborne Observatory. The rings of Uranus were discovered in 1977 using the 36-inch (91.5-centimeter) telescope aboard this converted C-141 jet transport. By flying high above the clouds, astronomers can study infrared radiation from celestial objects. These wavelengths are absorbed by water in the air before they reach the ground.
NASA

joked about the possibility of a ring system immediately but dismissed the idea because narrow rings were unknown and thought to be implausible. Elliot hit upon the correct interpretation four days later, soon after he returned to the United States. No other conclusion was possible: Uranus had a ring system.[3] Saturn was not unique.

Careful analysis of the flickers showed nine rather than five rings for Uranus—all very thin and dark compared with the rings of Saturn observed from Earth. It was the first ring system to be discovered for a planet in more than 350 years.

But how could such narrow rings be preserved? Collisions between particles in the rings should spread them out into a broad, flat band of debris like the rings of Saturn. So, typical for science, a major new discovery raised challenging new questions. How and when had such a strange ring system developed around Uranus? And how could it persist?

To Uranus and Neptune, Maybe

By the time that the rings of Uranus were discovered, two *Voyager* spacecraft were already at the Kennedy Space Center being prepared for launch. The prime objectives of the *Voyagers* were Jupiter and Saturn. Each probe would fly past both planets. *Voyager 2* could then be sent on to Uranus and Neptune, but only if *Voyager 1* succeeded at Saturn. If *Voyager 1* failed, *Voyager 2* would be retargeted for a look at Saturn's moon Titan, a satellite with an atmosphere. Even if *Voyager 1* succeeded and *Voyager 2* could travel on to Uranus and Neptune, there was no assurance that the spacecraft could continue operating long enough.

Yet if *Voyager 2* could survive, it could collect in six hours close to Uranus far more information about the planet than had been possible to gather in the 200 years since it was discovered. The spacecraft could:

- Measure more exactly the size and mass of Uranus.
- Analyze the atmosphere of Uranus.
- Detect a magnetic field of Uranus.
- Examine the known moons and rings of Uranus and look for undiscovered ones.
- Search for clues to explain the axial tilt of Uranus, the origin of its rings, and the evolution of the Uranian system.

A Grand Tour

On August 20, 1977, America's *Voyager 2* space probe rode a Titan-Centaur rocket away from Earth on a trajectory that could carry it by Jupiter in such a way that the largest planet's gravity would sling the craft on to Saturn. With care, it would have the right path so that the gravity of Saturn could, in turn, hurl the vehicle on toward Uranus. Again, with just the right position and time of encounter, *Voyager* could use the gravity of Uranus to venture on to Neptune. It was a Grand Tour of the four largest planets in the solar system.

Only once every 175 years do the outer planets align themselves so that a spacecraft can, with minimal fuel, use the gravity of one to change its course to another and repeat that process twice more so that four successively more distant planets can be reached. Through most of two centuries, a rocket leaving Earth can visit only one or occasionally two outer planets. But for a very brief period once in every 175 years, space engineers and scientists can squeeze four planetary visits out of one space probe—if that craft can last more than 12 years and still function in an environment where diminishing sunlight plunges the temperature toward 400° below zero Fahrenheit (−240° Celsius).

The *Voyager 2* Grand Tour mission to four planets almost didn't happen—three times.

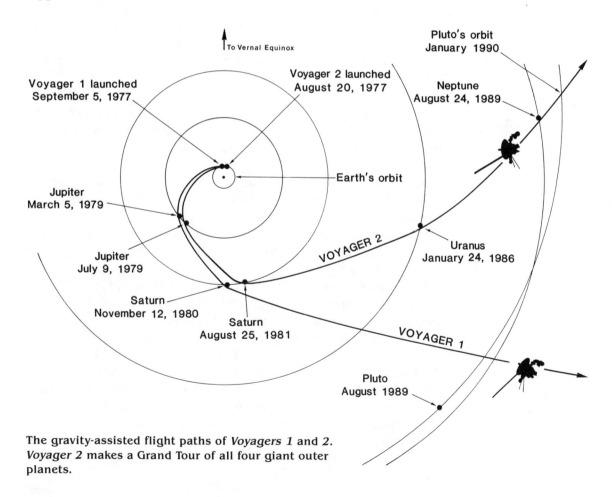

The gravity-assisted flight paths of *Voyagers 1* and *2*. *Voyager 2* makes a Grand Tour of all four giant outer planets.

The opportunity began to take shape in 1965 with discoveries by Gary Flandro and Michael Minovich that the gravity of the giant planets could be used to change the speed and course of a spacecraft so that it could fly on to a new destination. Specifically, Flandro determined, a spacecraft could use the gravity of Jupiter to accelerate on to an outer world with little or no further expenditure of fuel and with a great savings in time.

Most fascinating of all, Flandro found that the outer planets would be positioned in 1977 for a four-planet trip. A standard rocket journey direct from Earth to Neptune without gravitational assist would have required 40 years.[4] The Grand Tour, as Flandro named it, would allow a spacecraft to visit the four giant planets in 12 years.[5]

But there was a severe constraint. The last time the four giant planets had been favorably positioned for a Grand Tour was during the presidency of Thomas Jefferson. The next opportunity centered on 1977.[6] Could a spacecraft be built in time to take advantage of this rare planetary lineup? The challenges were formidable. In 1965, man had not yet walked on the Moon. Only one spacecraft, NASA's *Mariner 4,* had reached as far as Mars. To fly a 12-year mission so far from the Sun would not just stretch capabilities, it would require the development of new technologies.

Obstacles

With substantial power needs and yet sparse sunlight in the outer solar system, it would not be possible to use solar panels to provide electricity for the spacecraft. Instead the project would require miniature nuclear generators.

At such great distances from Earth, the spacecraft's onboard computer would have to be able to recognize and handle emergencies, since radio communication with the Earth would require hours.

Finally, the spacecraft systems and scientific instruments would have to last at least 12 years—a duration of time in 1965 that was longer than the space age was old.

As daunting as the challenge of the Grand Tour was, the problem of congressional approval was greater. The mission was canceled in December 1971. Most of the technical problems of a Grand Tour had been solved—Pluto had even been added to the itinerary—but Congress balked at a price tag of nearly $1 billion. The United States had reached the Moon but had no long-range plan for space exploration.[7]

In the summer of 1972, Congress voted $360 million for a less ambitious project: a flight to Jupiter and Saturn by two probes patterned after the Mariner probes that had successfully flown past Mars and Venus. The Jet Propulsion Laboratory in Pasadena, California, operated for NASA by the California Institute of Technology, would be the lead center for the project.

The shorter two-planet flight required spacecraft designed for only a 4-year mission—not 12. The *Mariner Jupiter-Saturn* project would use a gravitational assist at Jupiter to reach Saturn, but the Grand Tour mission to encompass four planets was dead.

DISCOVERY OF THE GRAND TOUR *VOYAGER* MISSION PROFILE
by Dr. Gary A. Flandro

PART 1

The work that led directly to the Grand Tour began at the Jet Propulsion Laboratory in 1965 when I was given a summer position to supplement my Ph.D. stipend as a graduate student in aeronautics at the California Institute of Technology. I had worked at JPL several times previously doing engineering work on missile trajectories, aerodynamics, and guidance systems. I learned as much from working there as from any of my formal graduate studies. The outstanding engineers at JPL had been a major inspiration for me to pursue graduate studies in space science.

My supervisor was Elliot "Joe" Cutting, with whom I had worked earlier on some trajectory problems. Joe assigned me the task of identifying possible unmanned missions to the outer planets. That was quite a leap at a time when America's longest spaceflight had been *Mariner 4* to Mars. The mere thought of missions to Saturn and beyond caused spacecraft engineers to tremble. The great distances to those bodies required long flight times. In 1965 the problem of building reliable mechanical and electronic devices with lifetimes long enough for trips to Mars (about 9 months) had not been truly solved. Missions that required vehicles to perform flawlessly for 9 years or longer were thought to be beyond our technical capability. Flights to Jupiter would take about 2 years, possibly just within our grasp, but missions to Neptune or Pluto would require approximately 40 years with the minimum energy transfer trajectories used in most space exploration. Another very worrisome problem was the difficulty of communicating over such vast distances. In light of these and other practical considerations, NASA and JPL management

Gary A. Flandro
Courtesy of Gary A. Flandro

had little interest in outer-planet exploration in 1965.

It was a great challenge to try to make exploration of the outer planets practical. I examined the conventional spaceflight trajectories to reach an outer planet with the least energy expended, in which a spacecraft is treated as a miniature planet in an elliptical orbit around the Sun. The vehicle's perihelion is the Earth's orbit, its aphelion is the target planet's orbit, and the flight is timed so that the spacecraft will arrive at the target planet's orbit just as the planet itself reaches that position. For trips beyond Jupiter, the flights took too long. We needed more speed, but we could accelerate the payload in a major way only during the rocket burn following launch. After that, a spacecraft bound for the outer planets constantly loses speed because of the Sun's gravity. To get more launch energy required either larger and more expensive rockets or much smaller and lighter space vehicles. Practical limits in both those directions had apparently already been reached. Was there some other energy source that could be tapped en route to increase the speed of the spacecraft? That was the key realization.

Astronomers had known since the late 1600s that when a comet passes close to a massive planet like Jupiter, its kinetic energy is changed tremendously and its orbit is greatly perturbed. Spaceflight pioneers understood this but did not realize its potential. The earliest study of "indirect" trajectories that used intermediate planets to mold the flight path in a desirable way was by Walter Hohmann in his book *Die Erreichbarkeit der Himmelskorper* (The Accessibility of the Celestial Bodies), published in 1925. He called these multiplanet trajectories the "Hohmann route" and designed the first Earth-Mars-Venus-Mercury flight paths.

In this work also, he first described the Hohmann minimum energy transfer orbit—the cost-effective trajectory utilized by a majority of planetary space missions.

Much later, indirect trajectories were proposed by Gaetano Arturo Crocco, the Italian scientist and aviation pioneer. He discovered that flight paths between the Earth, Mars, and Venus could be designed to utilize energy losses and gains in repeated close flybys to keep a space vehicle continuously in what he called the Grand Tour of the inner solar system. Crocco described his discovery in 1956 to the Seventh International Astronautical Congress in Rome.

The space age began the next year, but scant attention was paid at first to such trajectories in the technical literature on spaceflight. An exception was Krafft Ehricke, one of the original Peenemunde scientists. In *Space Flight* (1962), his voluminous work on applied celestial mechanics, he described the physical situation most clearly: "One rule, however, remains generally valid: *If at all possible, maneuvers for changing the heliocentric orbital elements should be carried out during the hyperbolic encounter with a planet, rather than in heliocentric space*. The greater the planet's mass, the greater the energy saving." But attention in the early 1960s was focused on the completion of simple one-target missions, so multiplanet flight paths did not attract much attention.

By 1965, however, JPL investigators were examining gravity-assist flight paths. Joe Cutting and Francis Sturms had devised a trajectory to the innermost planet Mercury that used a flyby at Venus to drop the spacecraft in toward the Sun. This concept became *Mariner 10*, the first successful multiplanet mission, and returned fantastic pictures of Mercury's surface.

Also working on gravity-assist possibilities that summer at JPL was Michael Minovich, a UCLA graduate student in astronomy, but our paths seldom crossed because he preferred to work at night. He was studying trajectories that used close flybys of Jupiter for the purpose of either escaping the solar system or making close approaches to the Sun. If a spacecraft caught up with a planet from behind, it gained energy and was flung outward at an increased speed instead of returning to the inner solar system on its original elliptical trajectory. If a spacecraft crossed in front of a planet, it lost energy and fell in closer to the Sun. Joe Cutting suggested that I examine gravity assist as a means for reaching the outer planets.

Strange as it now seems, many JPL engineers had misconceptions about gravity assist in 1965. They knew that because of gravity a spacecraft would gain speed as it approached a planet and lose speed as it coasted away, but they thought that there would be no net change in the spacecraft's energy relative to the Sun. They failed to consider that the planet was in motion around the Sun and could itself lose energy as it accelerated the spacecraft.

Another misconception was that multiplanet trajectories took a spacecraft out of the most direct orbital course to its final target and that therefore a gravity assist would *increase* rather than *decrease* flight time. This was a natural conclusion reached from examination of Hohmann's calculations and early multiplanet concepts such as those used in

Mariner 10. For *Mariner 10* to travel from the Earth inward toward the Sun required that kinetic energy be *reduced.* The net result was that flight time to Mercury via Venus was longer than a direct elliptical transfer would require.

The work of Minovich on solar system escape trajectories demonstrated that these were truly misconceptions. It became clear to me that the key to the outer solar system was to utilize the gravity-assist method. It was also obvious that Jupiter, with its enormous mass to bend spacecraft trajectories, was the best energy supply station, since its distance was reasonable, requiring typically a two-year flight time.

With these considerations in mind, I began detailed studies of Earth-Jupiter-Saturn trajectories. Previous work of this sort had been elementary, aimed just at establishing the feasibility of such orbits. My task was to calculate realistic mission profiles so that estimates of actual flight times, payloads, and planetary approach distances and speeds could be made. Of greatest importance was to identify "launch windows," periods during which such missions could be initiated.

In July 1965, I found that the best launch dates for a Jupiter-Saturn trajectory occurred in the late 1970s, perfect timing for the developing space-exploration program. I located the optimum launch dates by drawing graphs of the planetary longitudes for all of the outer planets.

It was at this time that I discovered something that had apparently not been noticed earlier: In the early 1980s, all of the outer planets would be on the same side of the Sun and in amazingly close proximity. This conjunction of the outer planets provided the inspiration for the Grand Tour mission concept. I could see immediately that a single spacecraft could explore all four giant outer planets by using each planet in succession to modify the spacecraft's trajectory as necessary to rendezvous with the next planet in the series.

PART 2

This was a rare moment of great exhilaration. Instantly it was mixed with a considerable amount of skepticism; I found myself doubting that anything practical could be done with the Grand Tour in view of our nation's slow progress in attaining spaceflight capability. However, ten years were available to overcome the engineering difficulties, and on second thought, motivation supplied by a goal like this one could have a real impact on progress.

I immediately began work to determine if practical multi-outer planet trajectories could be located. The trajectory computer programs available were not truly adequate for the job. I evolved a hand method using tabulations and graphs for "matching" the trajectories across each planetary encounter. Later, conic trajectory programs were developed that automated this tedious process. I set up a sequence of about ten trajectory runs each night and submitted these to the programmer (it was a job-shop computer operation in those days) and picked up the results the next morning. I would then examine these results to determine the next set of runs.

It took about a thousand trajectories to map out the original mission profiles and launch dates for the Grand Tour. These were plotted on graphs showing launch dates versus arrival dates, with spacecraft launch energy as the variable. This made it easy to visualize when and how the best mission possibilities would occur. The best launch window was in September 1978, with acceptable windows in 1977 and 1979. The actual launch dates used in the *Voyager* missions were virtually the same as those worked out by primitive methods in 1965.

Convincing others that the Grand Tour constituted a real mission opportunity was the most difficult part of the job. Cutting saw the possibilities immediately, but there were many naysayers in the ranks at JPL. They scoffed at designing a space vehicle that could survive many years and a close passage at Jupiter.

Eventually I was asked to present the Grand Tour concept to Homer Joe Stewart, one of my professors at Caltech, who also worked at JPL as director of the advanced concepts group. He saw the potential instantly.

The very next day, JPL issued a press release

describing the Grand Tour to the outer planets. Serious consideration of a multi-outer planet mission at JPL had begun. It eventually culminated in the successful *Pioneer 11* and *Voyager 1* and *2* missions.

Public opposition to the Grand Tour appeared at once. A hippie group, upon reading in the press release that the energy for flinging the spacecraft outward came at the expense of Jupiter's orbital energy, decided to organize the Pasadena Society for the Preservation of Jupiter's Orbit. They paraded in downtown Pasadena carrying signs, one of them wearing a flowing black cape and top hat, and held meetings for a short time in a good-natured way. But the real problems for the mission came later.

PART 3

The summer was over. My direct involvement with Grand Tour planning was finished, although for several years I continued to aid in the marketing of the mission concept by presenting technical papers and answering questions from the press. There was widespread acceptance of the idea with such notables as Wernher von Braun and President Nixon indicating support. In 1972, I received the annual Golovine Award of the British Interplanetary Society in recognition of my work in celestial mechanics. I had been nominated for this award by William H. Pickering, the JPL director.[1]

_____
[1]William Hayward Pickering (b. 1910) is no relation to William Henry Pickering (1858–1938), predictor of trans-Neptunian planets.

In the meantime, things had not gone well for the Grand Tour mission. The original very ambitious JPL plan involved a spacecraft that would have ejected atmospheric probes and orbiters at each intermediate planet. That mission was canceled in 1971 because of NASA budget restraints, but then a less ambitious *Mariner*-class spacecraft design was substituted. This "plain vanilla" plan was still based on the original Earth-Jupiter-Saturn gravity-assist flight path.

When it became apparent that the outer-planet Grand Tour mission would indeed fly, several individuals came forth claiming to be the originators of the idea. Truly, the multi-outer planet mission concept is the outcome of the work of many people. Walter Hohmann, originator of the multiplanet trajectory, and G. A. Crocco deserve as much credit as anyone for suggesting the use of gravity assist in planetary mission design and for the name Grand Tour. The solar system escape and close solar probe study by Mike Minovich demonstrated the benefit of the Jupiter gravity-assist maneuver.

The two *Voyager* spacecraft proved so reliable in the first legs of their flights that the *Voyager 2* flight plan was extended to perform the full four-planet Grand Tour mission, which will culminate when *Voyager 2* makes its final encounter, at Neptune, in August 1989, 24 years after my memorable summer at JPL.

Resurrection

Almost, but not quite. NASA began plans for a third *Mariner* probe to be launched on a Jupiter-Uranus mission in 1979. In 1975, the Jupiter-Uranus project and the third spacecraft were canceled for lack of funds.

In 1977, the *Mariner Jupiter-Saturn* project was given a more manageable name—*Voyager*. And the *Voyager* management team took a final fond look at the Grand Tour plan. The highest priority objectives at Jupiter and Saturn included not just data from the planets but also close-up pictures and measurements of Jupiter's innermost large moon, Io, and Saturn's largest moon, Titan.

It was still possible, with a 1977 launch, to fly the Grand Tour but only by sacrificing optimum views of Io and Titan. If *Voyager 1* flew suitably close to Io, it would acquire so much velocity from Jupiter that it would arrive at Saturn too soon to be diverted on to Uranus. If *Voyager 1* flew suitably close to Titan, it would leave Saturn headed in the wrong direction to reach Uranus.

But if *Voyager 1* gave up its option for a Grand Tour, it could examine both Io and Titan close up as the central mission objectives required. That would leave *Voyager 2* in a backup position—ready to undergo mid-course correction to fill in for *Voyager 1* if it failed at Saturn. But if *Voyager 1* succeeded at Saturn and Titan, *Voyager 2* could follow up and broaden the coverage at Saturn as it would at Jupiter while maintaining course for a Grand Tour mission on to Uranus and Neptune.

The flight to Uranus and Neptune, already twice canceled, was on again. It required no extra money, at least not immediately. It was only a contingency plan. A hope. A dream.

Voyager 2 was launched first, on August 24, 1977. *Voyager 1* lifted off on September 5, 1977, on a slightly more direct and hence faster route so that it would reach Jupiter and Saturn ahead of its sister ship. But project managers still considered a successful Grand Tour a long shot.

Voyager 1 succeeded brilliantly at Jupiter. It documented in detail the turbulent flow of the Jovian atmosphere, especially near the Great Red Spot, a centuries-old storm that is at least twice the size of Earth. *Voyager 1* found that Jupiter has a faint dusty ring.

Winds flow turbulently around the Great Red Spot on Jupiter, as seen by the *Voyager* spacecraft
NASA/Jet Propulsion Laboratory

Exploring Jupiter's moons, *Voyager 1* discovered active volcanoes on Io, the surface of Europa covered by cracked ice, parallel ridges and grooves on Ganymede, and a surface on Callisto so heavily cratered that new impacts erase as many craters as are formed.

A Grand Tour for *Voyager 2* was still a possibility.

Voyager 1 again performed splendidly at Saturn. It learned that the fabled ring system was composed not of a few but of thousands of rings. Winds in Saturn's atmosphere reached speeds of 1,000 miles per hour (1,600 kilometers per hour).

Voyager 1 discovered that Jupiter has a faint, dusty ring. The tiny particles showed up best as Voyager looked back through the ring toward the Sun to see sunlight scattered by the dust.
NASA/Jet Propulsion Laboratory

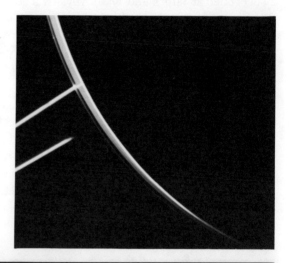

Voyager 1 discovered that Io, Jupiter's innermost large moon, has more active volcanoes than any other body in the solar system. A volcano on the horizon hurls ejecta more than a hundred miles above Io's surface.
NASA/Jet Propulsion Laboratory

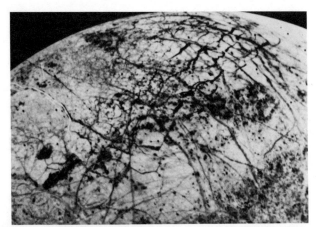

Jupiter's moon Europa has the smoothest surface of any body in the solar system. The cracks in its water ice crust do not jut upward to appreciable height. *NASA/Jet Propulsion Laboratory*

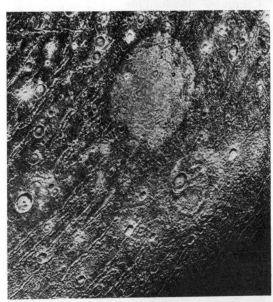

Jupiter's Ganymede, the largest moon in the solar system, has a highly wrinkled surface that suggests extensive movement of crustal plates. *NASA/Jet Propulsion Laboratory*

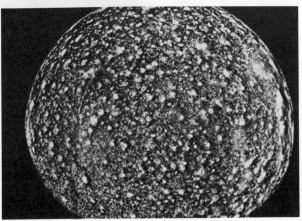

This high-resolution mosaic of Callisto, Jupiter's outermost large moon, shows that it is saturated with craters. Except for impact cratering, little has happened to Callisto since it formed. *NASA/Jet Propulsion Laboratory*

The *Voyagers* found that Saturn's ring system was composed of thousands of rings.
NASA/Jet Propulsion Laboratory

Computer-enhanced image of Saturn's atmosphere as seen by the *Voyagers.* The spots are smaller versions of Jupiter's Great Red Spot, but on Saturn the jet-stream winds flow three times as fast.
NASA/Jet Propulsion Laboratory

The *Voyagers* revealed for the first time surface details on the moons of Saturn. This is Dione.
NASA/Jet Propulsion Laboratory

Among Saturn's moons, Titan was known to have an atomosphere. *Voyager 1* found that this atmosphere was denser than Earth's and that its principal constituent was nitrogen. It revealed for the first time details on the other moons of Saturn, a wild variety of terrains with abundant evidence that, despite their modest size, these worlds had experienced internal evolution long after formation. *Voyager 1* discovered two new rings at Saturn and confirmed the theory that a narrow ring could be stabilized by small shepherd moons.

The Grand Tour for *Voyager 2* now loomed large.

On August 25, 1981, *Voyager 2* successfully cleared Saturn on course. The Grand Tour was on.

On—but again threatened, this time by problems with the spacecraft.

CHAPTER 8

The Grand Tour of *Voyager 2*

"The space program has done more to stimulate the growth of human knowledge than any other aspect of life on Earth."

Ellis D. Miner, Assistant Project Scientist for *Voyager*

Voyager 2 is legally deaf, chronically arthritic, and just a touch senile.[1]

A few months after *Voyager 2* set out for Jupiter, one of its radio receivers, vital for the spacecraft to receive instructions from Earth, failed outright and the backup receiver lost its tuning ability.

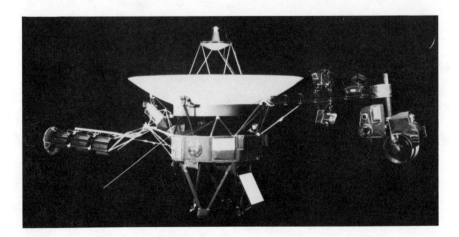

Voyager spacecraft
NASA/Jet Propulsion Laboratory

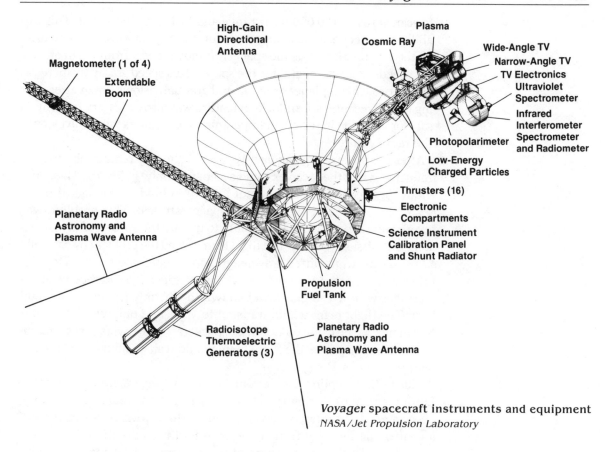

Magnetometer (1 of 4)

Extendable Boom

High-Gain Directional Antenna

Cosmic Ray

Plasma

Wide-Angle TV

Narrow-Angle TV

TV Electronics

Ultraviolet Spectrometer

Infrared Interferometer Spectrometer and Radiometer

Photopolarimeter

Low-Energy Charged Particles

Planetary Radio Astronomy and Plasma Wave Antenna

Thrusters (16)

Electronic Compartments

Science Instrument Calibration Panel and Shunt Radiator

Propulsion Fuel Tank

Radioisotope Thermoelectric Generators (3)

Planetary Radio Astronomy and Plasma Wave Antenna

Voyager spacecraft instruments and equipment
NASA/Jet Propulsion Laboratory

As it flew by Saturn, the *Voyager 2* scan platform, which turns the cameras and some of the other scientific instruments toward their targets, seized up.

Throughout most of the mission outward, a small portion of the spacecraft's computer memory has been inoperable.

Yet by the time it reached Uranus on January 24, 1986, after a flight of eight and a half years, *Voyager 2* was a better spacecraft than when it left home. The Jet Propulsion Laboratory scientific and engineering team had redesigned the spacecraft in flight.

Seven and a half months after launch, en route to Jupiter, a power surge on *Voyager 2* permanently destroyed the primary radio receiver and damaged the backup so badly that it could not tune to receive different frequencies from Earth. Before the failure the receiver had been able

Almost Showstoppers

to scan across a 100,000-hertz range and lock onto the signal. This capability was necessary to compensate for the varying wavelengths caused by the Doppler Effect as the spacecraft moves away from Earth at varying speeds. Because of gravity, the spacecraft's speed gradually increases as it approaches a planet or moon and gradually slows down after passing it. The receiver's scan/lock-on feature was also necessary to compensate for variation in receiver temperature, which alters what wavelengths the radio can receive.

After the power surge aboard *Voyager 2,* the tiny remaining range of frequencies that the receiver could accept was only ±96 hertz. The bandwidth for reception had been cut a thousandfold. A change of 0.5°F (0.25°C) was enough to render the spacecraft deaf. This problem was severe and scary. Without the ability to update the spacecraft's instructions, the mission was lost. The flight team solved the problem by carefully adjusting the transmitted wavelengths to compensate for the Doppler Effect and expected temperature so that *Voyager 2* could receive all its messages within a single small bandwidth, the only frequency it could hear. The flight team would transmit to *Voyager 2* only when the craft was in normal cruising mode. After spacecraft maneuvers that generate heat within the probe, the flight team would wait 48 hours before sending messages to *Voyager 2.*

This radio reception problem meant that *Voyager 2* would be deaf to Earth commands during the close phases of the Uranus and Neptune encounters. No matter. A message from Earth to *Voyager* at Uranus and a confirming message from *Voyager* to Earth required five and a half hours—almost the entire length of the *Voyager 2* close encounter with Uranus. The spacecraft would have to be very carefully preprogrammed with instructions for the entire close-in data-gathering and picture-taking sequence at Uranus and Neptune in order to enjoy a successful flyby. By the time some instructional error was revealed during the encounter, there would be no time to correct it. And there would be no second chance.

And just in case this final frequency fails, the flight team always keeps *Voyager 2* loaded with the latest instructions to carry out the Neptune flyby without further update from Earth.

The technical challenge of getting good information from Uranus was intimidating. Uranus lies at twice the distance of Saturn, so there is only one-fourth as much sunlight at Uranus to provide illumination for pictures. Thus long exposures are necessary. But long exposures of moving objects lead to smeared images unless the camera is moved to track the target.

Unfortunately, the scan platform on which *Voyager 2*'s cameras were mounted had jammed as it pivoted rapidly to follow targets one and a half hours after its closest approach to Saturn on August 25, 1981.[2] A number of pictures and measurements scheduled to be made were lost because the spacecraft could not point its instruments in the right direction. Lubrication gradually seeped back into the gears and the system seemed to be working again at slow and moderate speeds, but the mission team wanted to take no chances. The Uranus encounter would be so brief that even one picture was too precious to be lost. So whenever fast or even moderately fast slews were required of the scan platform to track the targets, the entire spacecraft was rolled to follow the action.[3] Rolling the entire craft provided smoother motion besides. But it was tricky. *Voyager 2* would rush by the moon Miranda at 20 times the speed of a rifle bullet while the whole spacecraft rolled to pan its cameras to an accuracy of better than one-tenth of a degree.

The enhancement of *Voyager* to meet the challenge of Uranus and Neptune did not stop with an understanding of and a compensation for the scan-platform problem. In the weak sunlight at Uranus and with *Voyager* moving as fast as 56,000 miles per hour (25 kilometers per second), additional care had to be taken to make the craft as steady as possible during picture taking. Even the starting and stopping of *Voyager*'s tape recorder was a problem. Every time the recorder turned on, the start-up of the spinning reels created equal and opposite momentum in the spacecraft. The spin induced in *Voyager* was very tiny because of its much greater mass, but the spin was nevertheless very real and could blur pictures and foul up other directional measurements. When the tape recorder stopped, the spacecraft picked up that momentum and began to spin slowly in the direction that the reels had been turning. For every action, there is an equal and opposite reaction. It was a textbook demonstration of Newton's third law of motion. It was also a troublesome obstacle to quality pictures and other measurements expected from *Voyager 2*. The mission team solved the problem by programming *Voyager 2* to fire tiny bursts of gas from its thrusters to counteract the torque created by the starting and stopping of the onboard tape recorder.

Voyager 2 also ran afoul of a computer memory problem when a computer chip failed in one of the two scientific computers. The damage was irreparable and cost that computer 3 percent of its memory capacity. For this problem, there was no "fix" except to moderate demands on the computer.[4]

Thus, on its way to Uranus, *Voyager 2* developed three potentially showstopping technical problems—with radio reception, with its scan platform, and with its computer memory. From Earth, millions of miles away, these problems had been detected, diagnosed, and worked around so that *Voyager 2* lost very little of the capabilities built into it and did not have to forgo any of the experiments planned for it to execute.

But the ground team went much further than restoring the spacecraft to its initial proficiency. They actually made it better.

Enhancements

As *Voyager 2* sped farther from Earth, its radio signal became weaker. At Jupiter, the signal was weak but adequate. At Saturn, almost twice as far away, the signal was four times weaker. At Uranus, more than twice the distance of Saturn, the signal upon which all the data would be received on Earth was four times weaker than from Saturn and sixteen times weaker than from Jupiter. The result was that in order to avoid a jumbled message, *Voyager 2* could not pack its data so tightly for transmission, just as a speaker talks more slowly to make certain he is understood.

Especially critical were the images to be transmitted, which required vast amounts of transmission time. All *Voyager* pictures are black-and-white, but taking multiple pictures through different filters allows scientists back on Earth to combine filtered pictures to make color images. Even so, transmitting black-and-white pictures requires enormous quantities of bits. Each *Voyager* image is a television picture formed from an array of 800 pixels (picture elements, dots) across and 800 pixels down—a total of 640,000 pixels. And the darkness or lightness of each of these pixels is described by 8 binary bits to provide a gray scale of 256 shades. Thus each picture to be transmitted required $640,000 \times 8 = 5,120,000$ bits of information—over 5 million bits.

At Jupiter, *Voyager*'s X-band transmitter could start and finish broadcasting a picture to Earth within 48 seconds.[5] At Saturn, where distance forced *Voyager 2* to use a lower data transmission rate, the time between the start and finish of a single picture broadcast was about two and a half minutes. At Uranus, the transmission time for a single image would rise to ten minutes—unacceptably long. When *Voyager* rolled so that its cameras could track a target, its antenna would be turned away from Earth and its tape recorder would store the images until they could be transmitted. With transmission times so long, there was danger that the recorder would be quickly overwhelmed and precious pictures and other data would be lost.

The computer scientists at the Jet Propulsion Laboratory came up with a solution—a computer innovation called data compression. Instead of describing the exact shade of gray of each pixel in an image, only the first pixel in each line would be precisely described. Each pixel that followed in that row would then be described only in terms of how much lighter or darker it was than the one preceding it. This data compression carried risks. An error (due to electronic noise) at the beginning of a row would create errors in all the other pixels in that row. And an error in any pixel in the row would make the values of all following pixels in that row wrong. The image as a whole, however, ought to reveal such defects by contrast.

There was another risk. Data compression would have to be performed by the computers aboard *Voyager 2*, and the computer program to compress the data was so extensive that it would require one of *Voyager*'s two scientific computers to be completely dedicated to this function alone.

Yet it was worth the risk. The data-compression system reduced the number of bits required to send an image by 60 percent. Instead of 12 minutes to broadcast one picture from Uranus, it would now take only 5.

But data compression was not the only improvement to *Voyager 2* made en route to Uranus. *Voyager* engineers also worked out a new system for shorter bursts on the thrusters to stabilize the craft, making it smoother in its maneuvers. *Voyager 2* became able to take steadier pictures while becoming more conservative of fuel.

So as *Voyager 2* got older, it actually did get better, thanks to the ingenuity of a lot of scientists and engineers on Earth.

Bigger Ears

And facilities on Earth improved as well. They had to if they were going to hear the puny radio signal of *Voyager* all the way from Uranus. "Clear-channel" radio stations in the United States broadcast on 50,000 watts of power. *Voyager 2* transmits its data to Earth on 22 watts. But that signal spreads out during its journey from Uranus so that by the time it has traveled 1.75 billion miles (2.8 million kilometers), a 210-foot (64-meter) dish receiver on Earth captures only one-billionth of one-millionth of a watt.

NASA's large Deep Space Network radio receiving dish at Goldstone, California
NASA/Jet Propulsion Laboratory

To receive *Voyager 2*'s information from Jupiter and Saturn and to transmit instructions to the spacecraft, NASA had used its Deep Space Network—three tracking stations scattered a third of the way around the world from one another. One is at Goldstone, California, 100 miles (160 kilometers) northeast of Los Angeles. The second is near Madrid, Spain. The third is near Canberra, Australia. Each Deep Space Network site was equipped with a steerable parabolic dish antenna 210 feet (64 meters) in diameter.

For the Uranus encounter those dishes weren't enough. NASA augmented their receiving capability by linking the giant dishes to smaller 112-foot (34-meter) dishes at each station. It worked like an old-fashioned ear horn for a partially deaf person: the more receiving area, the more signal captured. At Madrid, a single 112-foot antenna was arrayed with a 210-foot dish, boosting reception by about 28 percent.

The timing of the encounter placed the greatest burden on the Canberra and Goldstone sites. There, two 112-foot antennae were arrayed with the 210-foot dish, and receptivity was increased by about 56 percent. At Canberra, the receiving site for *Voyager*'s data from Uranus at closest encounter, a further enhancement was made. The Australian government linked its 210-foot (64-meter) Parkes radio telescope with the three-dish array of NASA's Deep Space Network at Canberra to form a combined antenna with the reception power of a 400-foot (100-meter) dish.

Bull's-eye

In addition to the chronic ailments of its remaining radio receiver, its scan platform, and its computer memory, and in addition to the greater distance that affected the strength of the radio signal from *Voyager 2* and the time for instructions to reach the spacecraft, *Voyager 2* faced a special challenge at Uranus because of the planet's orientation.

The two *Voyager* probes had previously encountered Jupiter and Saturn, planets that, like Earth, travel around the Sun with their rotational axes nearly perpendicular to their orbital planes. Because they travel around the Sun "standing up," their equators lie nearly in their orbital planes. All the rings and a great majority of the moons in the solar system revolve around their planets' equators. So the rings and most of the moons of Jupiter and Saturn lie nearly in the plane of revolution for that planet, which is extremely close to the orbital plane of the Earth and hence essentially the plane formed by flight of *Voyager 2*. Thus as *Voyager 1* and *2* approached Jupiter and Saturn, it was a little like approaching a city on a flat map. *Voyager* first found itself in the suburbs, able to shoot a snapshot of one moon and then prepare for the next. In sequence, the spacecraft passed one moon after another, then closest to the planet, and then outward past additional moons one at a time. The Jupiter and Saturn close encounters were each spread out over several days.

Uranus was not so accommodating. *Voyager 2*'s closest encounters with Uranus, its rings, and all of its moons were crammed into a period of less than six hours.

The rings and moons of Uranus revolve almost precisely around its equator, just as most of the moons and all of the rings of the other planets do. The problem is that Uranus lies on its side as it revolves around the Sun. So the orbits of its moons and rings are tilted, like the planet, almost perpendicularly to the orbital plane of Uranus. Even so, if the equator of Uranus had been more or less edge on to *Voyager* as it approached (which is approximately the case for about half of the 84-year period of Uranus), *Voyager* could have passed across that orbital plane and the moons could have been studied sequentially on approach and departure.

But the planetary alignment that allowed *Voyager 2* to bounce from

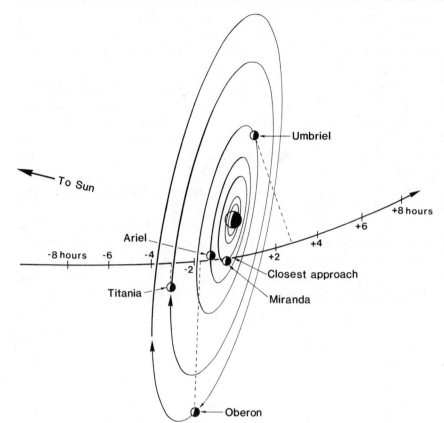

Umbriel

To Sun

-8 hours   -6   -4
-2
Ariel
Titania
Miranda
Closest approach
Oberon

+2   +4   +6   +8 hours

Voyager 2's **flight past Uranus.** *Voyager 2* **passed nearly perpendicularly through the Uranian system because Uranus lies on its side as it orbits the Sun and its moons orbit above its equator. At the time of** *Voyager 2's* **arrival, Uranus had its south pole pointed toward the spacecraft and the Sun and thus presented a bull's-eye target to its visitor. Because of this geometry,** *Voyager 2's* **closest approaches to Uranus, its rings, and each of its moons (its opportunities for best pictures and measurements) were squeezed into a period of less than six hours.**

Jupiter to Saturn to Uranus to Neptune just happened to catch Uranus with its south pole pointed toward the Sun, the Earth, and the approaching robot ambassador from Earth. *Voyager 2* saw Uranus, its rings, and its moons laid out before it like a target with a bull's-eye and concentric rings. The spacecraft plunged perpendicularly through the Uranian system like a dart thrown at a paper target. The time when *Voyager 2* was closest to Uranus was also very nearly the time when *Voyager 2* had its closest view of each moon. The picture taking and scientific measurements that could be spread over several days at Jupiter and Saturn were compressed into a little over five hours at Uranus. While each image was made, another important target was near its best viewing.

It had taken four and a half years for *Voyager 2* to fly from Saturn to Uranus. Working against formidable obstacles of distance, darkness, and planetary orientation at Uranus; working against a crippled radio receiver, a sticky scan platform, and a degraded computer onboard the spacecraft,

Solo

scientists meticulously planned every aspect of the Uranus encounter so that *Voyager 2* could take full advantage of the first practical opportunity since the dawn of the space age to examine in detail the realm of Uranus.

At Jupiter and Saturn, *Voyager 2* had been the backup craft, following up on *Voyager 1*'s discoveries and duplicating measurements from a different angle. Now, at Uranus, *Voyager 2* was the only spacecraft, traveling where no spacecraft had ever gone before. With each passing moment, *Voyager 2* was expanding the absolute frontier of planetary journeys.

On January 24, 1986, eight and a half years outbound from Earth, NASA's *Voyager 2* spacecraft passed 50,700 miles (81,600 kilometers) above the cloud tops of Uranus. Uranus was 1.75 billion miles (2.8 billion kilometers) from Earth, and the data radioed from *Voyager* took two and three-quarters hours to reach home traveling at the speed of light.

Astronomy had come a long way in the 205 years since Herschel had caught sight of a tiny disk in his homemade telescope.

CHAPTER 9

Triumph at Uranus

"What do we learn about the Earth by studying the planets? Humility!"
Planetary scientist Andrew P. Ingersoll
(1986)

The blue-green planet that *Voyager 2* approached in January 1986 had clouds, but not in brilliant earth-tone colors like those of Jupiter. It had rings, but not in bright, broad belts made of thousands of ringlets like those of Saturn. It had moons, but none as large as Jupiter's Big Four or Saturn's Titan. Uranus was more subtle, yet every bit as puzzling. Its atmosphere seemed bland, its rings were narrow and dark, its largest moons were quite modest in size. It would take a connoisseur to appreciate this planet—and a great many imaginative and clear-thinking scientists to understand it.

The atmospheres of Jupiter, Saturn, Uranus, and Neptune are basically the same. It is their temperatures that make them look so very different. Their atmospheres are composed overwhelmingly of hydrogen and helium, with only traces of other chemicals. All have clouds made of some of those trace elements. All have an obscuring haze layer formed as sunlight breaks up methane molecules (CH_4), allowing the carbon to recombine into more complicated and less transparent hydrogen-carbon compounds, such as acetylene (C_2H_2) and ethane (C_2H_6).

Each of these gas giant planets has a well-mixed atmosphere rich in chemicals. On Earth, most of those chemicals are dissolved in the oceans

Atmospherics

or have been compounded to form rocks and hence have been extracted from the air. But on Jupiter, Saturn, Uranus, and Neptune, all those chemicals are gases in the atmosphere and will condense to form clouds if the temperature is right. On Earth, only water condenses and forms white clouds. On Jupiter, temperatures at various levels are cold enough to form ice clouds of water (white); ammonia (white); and compounds of ammonia (white), phosphorus (reddish brown), sulfur (red, orange, yellow, blue, and black), and various more complex organic compounds (usually reddish). At Jupiter, heat from the Sun and from deep within the planet creates updrafts in the atmosphere. These rising gases expand, cool, and then sink back to lower, warmer levels. As each gas in the atmosphere reaches its special temperature of condensation, it freezes out of the air to form clouds of ice particles—different colors for different chemicals.

Each of the gas giant planets produces clouds according to the heat it receives from the Sun and its own internal sources of heat. The colder a planet is, the less vertical atmospheric motion it has and the fewer cloud patterns and features.

But another factor is at work on the gas giants that determines what cloud features are visible when the planet is viewed from a distance. That factor is the hydrocarbon smog. This obscuring haze is present on all four gas giants at a level where the atmospheric pressure is about one-hundredth to one-tenth the pressure at sea level on Earth. In the comparative warmth on Jupiter, the clouds billow up above the smog to create a planet that looks like an abstract painting in riotous colors. At Saturn, there are still patterns, storms, and eddies to be seen but less distinctly

Using false color, clouds were detected in the Uranian atmosphere; one is visible here near the one o'clock position close to the planet's edge (near Uranus' equator). The small doughnut-shaped features are not real; they were caused by dust in *Voyager 2*'s camera.
NASA/Jet Propulsion Laboratory

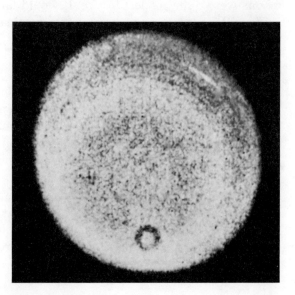

and less colorfully. Saturn's lower temperature creates less cloud activity, and much of it is buried out of sight beneath the smog.[1]

Over the decades that preceded the visit of *Voyager 2* to Uranus, several Earth-based astronomers had claimed to see clouds in its atmosphere, but their contemporaries had been unable to confirm these findings. *Voyager 2* did indeed find clouds drifting in the Uranian atmosphere, but these clouds were visible only through special filters and with contrast enhanced by computers. They are invisible at ordinary optical wavelengths because, with so little solar or internal heat at Uranus, the clouds lie deep in the atmosphere below the thin high haze of hydrocarbon smog that veils the planet. Below the haze is a cloud deck of tiny methane ice crystals two to three miles thick floating in the planet's predominantly hydrogen atmosphere.

The Weather on Uranus

Uranus lies on its side as it revolves around the Sun. How would this orientation affect the weather *Voyager 2* would find on the planet?

The tilt of Uranus means that during the course of a year on the planet—equal to 84 Earth years—the Sun stands directly or almost directly above every point on Uranus.[2] Thus, for several years out of each 84-year orbital cycle, the north and south poles on Uranus are treated to sunlight beaming down from near the top of the sky—a circumstance unique to Uranus and Pluto among planets in the solar system.

The orbit of Uranus causes the position of the Sun to appear to slowly swing from pole to pole during the course of a Uranian year. As the Sun

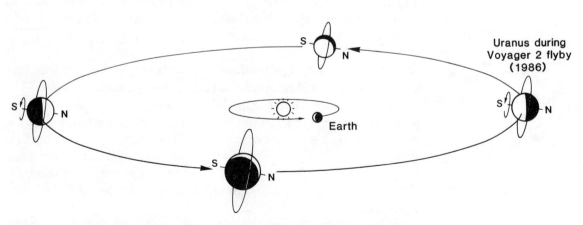

The seasons of Uranus. Uranus lies on its side as it orbits the Sun. This axial tilt means that the Sun can stand almost directly above the planet's north and south poles. For 42 years, the south pole receives continuous sunlight. Then, for the next 42 years at the south pole, the Sun is never visible.

is hovering nearly over each pole, the seasonal cycle of the Sun's north/south apparent motion in the sky is in the process of slowing, stopping, and reversing itself. At midcycle, when the Sun is over the equator, the seasons are changing fastest. Therefore, the polar regions of Uranus actually receive slightly more solar warmth than the equator. Not much, because the Sun provides precious little heat at Uranus, but scientists expected that the sunlit pole would be a few degrees warmer than the equator.

It wasn't. *Voyager 2* found that the temperatures around the planet, poles and equator, day and night, were within 3° Fahrenheit of being identical—about 366° below zero Fahrenheit (−221°C; 52°K). The sunlight is so weak that what little heat is received is quickly spread rather evenly around the planet by its winds.

But how would winds, if any, flow on a planet where the region around the south pole is currently receiving the most direct sunlight? On Earth, the Sun's radiation falls most vertically and hence most intensely, near the equator, where it heats the surface air and causes it to rise. Cooler air farther from the equator flows in to form the principal wind circulation. The west-to-east rotation of the Earth then causes those air currents to curve so that the prevailing winds blow west to east.

But with maximum heating on Uranus at the south pole, would the prevailing winds on that planet be north to south? No, *Voyager* found, the winds blow east to west because of the rapid east-to-west rotation of Uranus.[3]

Meterologists were startled. The angle of sunlight is less important than planetary rotation in generating the motion of weather systems. The jetstream winds in the upper atmosphere of Uranus flow at 200 miles per hour (300 kilometers per hour), about twice the speed of those on Earth.

The Magnetic Field of Uranus

Perhaps the greatest surprise afforded by the planet itself was its magnetic field. A magnetic field was expected because Uranus is very massive—14.5 times more massive than Earth. The mass of Uranus is great enough that its gravity can create a hot interior. This heat makes atoms shed electrons so that they are electrically charged. The rotation of the planet then spins this electrically charged fluid interior to generate a magnetic field around the planet. The magnetic field changes the courses of incoming charged particles from the Sun and ionized particles in the planet's atmosphere.

But from Earth it had not been possible to detect a magnetic field for Uranus. *Voyager 2* did the honors—and raised some eyebrows. The magnetic field of Uranus was greater than the Earth's. It was also absurdly tilted. The magnetic poles of Earth correspond roughly with the poles of rotation. They are offset from one another by only about 11.7 degrees. The same is true for Jupiter and Saturn. But not for Uranus. The magnetic axis of Uranus is tipped 60 degrees to the poles of rotation, meaning that the north and south magnetic poles lie closer to the equator than to the geographic poles.

But that wasn't all that was strange. The magnetic axis of the Earth passes very nearly through the exact center of our planet. (It misses by less than 300 miles [500 kilometers]). So it is at Jupiter and Saturn as well. But not at Uranus. There the magnetic axis lies 4,800 miles (7,700 kilometers) off center, passing through Uranus about one-third of the way between the planet's center and its surface.[4]

Geophysicists interpret these oddities to mean that the magnetic field of Uranus is generated not at or close to the planet's center but outside of the core.[5] The Earth's magnetic field is thought to come from molten iron and nickel. The magnetic field of Uranus is thought to come from its mantle where water and ammonia may be under sufficient pressure to make them good electrical conductors.

It may be that the 60-degree tilt of the magnetic field is just an accident in time. Perhaps *Voyager 2* reached Uranus while it was undergoing a magnetic field reversal in which the poles switch magnetic polarity, something that happens on Earth about once every half-million years.[6]

Voyager 2 used the planet's magnetic field to determine the length of a day on Uranus. As the magnetic field swept over the spacecraft, increasing in intensity whenever a magnetic pole rotated by *Voyager*, scientists could observe that the magnetic field of Uranus spins around once every 17.24 hours. For practical purposes, this cycle is the rotation period of the planet. Uranus has no solid surface and hence no landmarks that repeatedly rotate by to reveal the length of a day.

Spectroscopic measurements of the spin of Uranus from Earth could only approximate the rotational period. To calculate the rotation of Uranus based on cloud motions in the atmosphere was problematical because, as on Earth, wind speed varies with location and time. At present, the fastest winds on Uranus are near the south pole, where the solar heating of the atmosphere is most intense. A year before *Voyager 2*'s encounter, the rotational period of Uranus was variously listed between 10 and 24 hours, with newer estimates suggesting 15 to 17 hours. With its measurement of 17.24 hours, *Voyager 2* refined that figure considerably.

Uranus and Neptune rotate faster than Earth and the other rocky midget planets but not as rapidly as the larger giants, Jupiter and Saturn.

Dayglow

The spacecraft found all sorts of atmospheric glows at Uranus, just as expected based on what the *Voyagers* had detected at Jupiter and Saturn.

Uranus had an aurora, similar to the northern and southern lights on Earth. Subatomic particles from the Sun were diverted by the magnetic field of Uranus toward the planet's magnetic poles, where they plunged into the atmosphere, causing hydrogen molecules, the principal constituent, to glow.

Uranus also exhibited air glow on its daytime face, as ultraviolet energy from the Sun caused some atmospheric atoms and molecules to glow very faintly.

But most interesting was a previously unrecognized atmospheric emission initially given the name *clectroglow* and now renamed *dayglow* to emphasize that it happens only when the atmosphere is directly exposed to sunlight. Dayglow had been detected by the *Voyagers* at Jupiter and Saturn but initially misidentified as aurora. *Voyager 2* revealed it unmistakably at Uranus.[7] The exact cause of dayglow is still being debated, but the fluorescence seen comes from hydrogen atoms in the tenuous upper atmosphere only on the sunward side of the planet. The total ultraviolet light released by dayglow from the sunlit atmosphere of Uranus corresponds to a few trillion watts of power. One trillion watts is enough to power 20 million homes at peak demand—all the homes in the New York and Los Angeles metropolitan areas.[8]

The Energy of Uranus

Uranus was expected to be slightly warmer than heating by the distant Sun would allow. Jupiter and Saturn are radiating into space almost twice as much energy as they receive from the Sun.[9] They emit this energy not as visible light but in the form of heat and radio waves. In the case of Jupiter, the source of this radiation is energy released by the continuing decay of radioactive elements within the planet and from the heat as gravity continues to compress its interior. In the case of Saturn, the internal heat is mostly caused by radioactive decay and gravitational compression, but about a third of Saturn's excess radiation comes from the separating mix of hydrogen and helium within the planet. The heavier helium is falling toward the planet's core, creating heat by friction as it rubs against the hydrogen fluid.[10]

Although it is not the size of Jupiter and Saturn, Uranus is a giant planet, so gravitational and radioactive heating could be anticipated. *Voyager 2* did indeed find Uranus slightly warmer than sunlight alone could provide, but not much. Uranus emits only about 15 percent more energy than it receives from the Sun, energy radiated in the infrared and radio wavelengths.

Measurements by *Voyager 2* greatly clarified our perspective on the structure of Uranus. The planet is about 85 percent hydrogen and 12 percent helium by mass—close to the same proportions as the Sun, Jupiter, and Saturn. The bodies of the solar system are truly a family, spawned from the same cloud of gas and dust and differing from one another because of their distances from the Sun.

Hydrogen dominates the atmosphere of Uranus, but the visible "surface" of the planet is a haze layer of acetylene (C_2H_2) and ethane (C_2H_6) formed by sunlight striking methane in the Uranian atmosphere. It is the small amount of methane (2 percent to 3 percent) mixed in with the hydrogen, helium, and other elements and compounds that gives Uranus its blue-green color by absorbing and not reflecting the red portion of

The Structure of Uranus

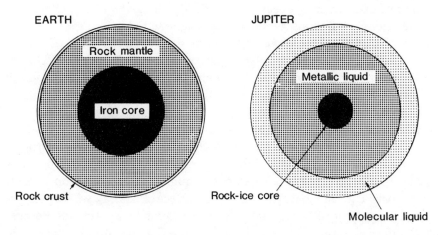

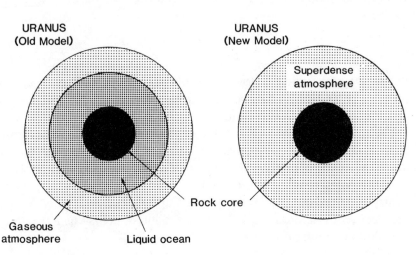

The internal structure of Uranus compared to that of Earth and Jupiter. *Voyager 2* indicated that a two-layer model for the interior of Uranus is correct rather than the earlier three-layer model. (The internal structure of each planet is drawn to scale, but not its size compared to the others. Uranus is about 4 times the diameter of Earth; Jupiter is more than 11.)

the sunlight it receives. The upper atmosphere of Uranus, warmed only feebly by sunlight and the weak internal heat of the planet, is extremely cold—about 350° below zero Fahrenheit (−212° Celsius).

Before *Voyager 2*'s visit to Uranus, the interior of the planet was thought to be at least basically similar to the interiors of Jupiter and Saturn—a hot, rocky core about the size of Earth surrounded by a dense ocean of hydrogen under so much pressure that it behaved like a metal. On top of this dense fluid, where the pressure was less, lay the gaseous atmosphere, mostly hydrogen and helium.

Initial analysis of *Voyager 2* data suggested that this three-layered structure applied to Uranus as well, except that the behavior of the mantle surrounding the core was dominated by liquid water mixed in with the hydrogen and under so much pressure that it was ionized. Because of its electrical charge, this water became a good conductor of electricity. As this giant interior ocean of electrically charged water rotated, said the scientists, Uranus generated its magnetic field.

But late in 1987 that conclusion was dramatically reversed.[11]

The problem is that for a planet that rotates as rapidly as it does, Uranus has relatively little bulge at its equator. Given its speed of rotation, if Uranus had a light atmosphere overlying a dense liquid mantle, the planet should bulge more as it turns. But it doesn't. Therefore, said the new conclusion, Uranus probably does not have a light atmosphere on top of a liquid mantle. Instead the gases of the atmosphere—hydrogen, helium, water, ammonia, and methane—are all mixed together as a superdense atmosphere—a gas, not a liquid—that extends from outside the rocky core all the way to the visible surface of Uranus, where pressure and temperature allow different molecules to condense as clouds. Because Neptune is similar in size and mass, its interior structure is expected to be the same.

The Problem of Tilt

In a solar system where all the planets out to Uranus stand nearly upright on their polar axes as they revolve around the Sun, why does Uranus lie on its side? *Voyager 2* found no evidence to resolve this puzzle.

The still-prevailing view is that Uranus and all the planets were formed from a vast cloud of gas and dust by accretion: They were hit by, and absorbed, many cometlike planetesimals of differing sizes. Some struck near the poles, tending to tilt the planets. In some planets the impacts nearly balanced out, leaving the planets upright as they formed. Others were left with moderate tilts. The succession of collisions left Uranus and Pluto spinning on their sides.

Before the arrival of *Voyager 2*, Uranus was known to harbor five moons. But Earth-bound telescopes had not been able to resolve the moons' disks, so their diameters could only be approximated. They were substantial in size but far smaller than the four largest satellites of Jupiter, which range from just slightly smaller than our Moon to just slightly larger than the planet Mercury. None was the size of Titan, the largest moon of Saturn, or Triton, the largest moon of Neptune.

Voyager 2 provided the first pictures ever made that showed features on these five previously recognized but little-known moons of Uranus. The spacecraft also discovered ten new satellites of Uranus and two more rings to add to the nine already known.

Uranus is about a thousand times more massive than all its satellites and its ring system put together—very roughly the same ratio as its fellow gas giant planets. But of the four giant planets in our solar system, Uranus is the only one without a giant moon. Even the Moon of Earth, which is 2,160 miles (3,476 kilometers) in diameter and ranks sixth in size among planetary satellites, is more than twice the diameter of the largest Uranian moon. If our Moon were hollow, it would be possible to fit into it all 15 known moons and all the particles in the 11 known rings of Uranus and still have with room left over.

Small they were—but not without many surprises.

The two biggest moons are the two outermost: Oberon and Titania. Almost the same in size, they are otherwise very different. Next, moving inward, are Umbriel and Ariel, almost identical in size and about three-quarters the diameter of Oberon and Titania. But again, size is about all they have in common. The innermost of the major moons is Miranda, smallest of the five—and the weirdest.

The first of the Uranian moons to be discovered had been Oberon and Titania, found on January 11, 1787, by William Herschel, about six years after he had discovered Uranus. This time he was using his 18.8-inch (20-foot focal-length) reflector. Umbriel and Ariel were discovered in 1851 by William Lassell, an English amateur astronomer in the tradition of Herschel.

The names for the moons of Uranus were provided about 1852 by John Herschel, William's son, who borrowed them from English literature rather than Greek and Roman mythology. A German astronomer protested this violation of tradition. The names Oberon and Titania were taken from Shakespeare's *A Midsummer Night's Dream*. Umbriel and Ariel were given names snipped from Alexander Pope's *The Rape of the Lock*.[12] For Miranda, not discovered until 1948 by Gerard P. Kuiper, the source of the name was once again Shakespeare, this time *The Tempest*.

Like most satellites elsewhere in the solar system, the moons of Uranus

The Moons of Uranus

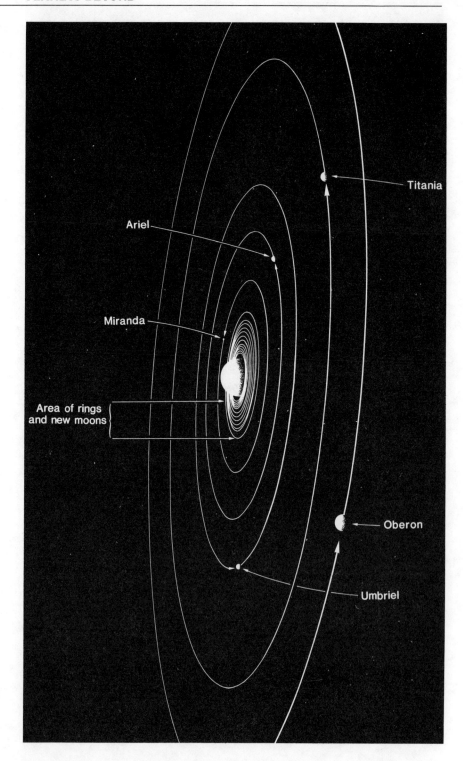

Uranus, its rings and
moons

were expected to have nearly circular orbits, to revolve around their planet's equator, and to rotate on their axes in the same amount of time as they revolved so that they always kept the same face toward Uranus (as our Moon always keeps the same face toward Earth). Despite the tilt of Uranus and its retrograde rotation, all 15 of its moons comply with these conventions.[13]

The moons of Uranus were also supposed to be dull. After all, they were small and cold—a good formula for dead worlds. Here were no moons like Saturn's Titan, so large that its gravity could retain an atmosphere. No moons like Jupiter's Io, which was still being pulled periodically like a piece of taffy between Jupiter and the moons Europa and Ganymede so that Io's insides are molten and it belches lava to the surface and into space with a frequency and ferocity unmatched by any other body in the solar system.

Based on all the bodies circling the Sun and especially the other satellites in the solar system, the moons of Uranus were expected to be simple little worlds of rock and ice that had formed along with or soon after Uranus, in time to be pounded by icy and rocky debris left over from the accretion of the planets and the moons. This bombardment should have saturated their surfaces with large and small craters—craters overlapping craters. For moons as small as those of Uranus, it was expected that little had happened since their formation except the impact of an occasional comet onto a surface covered with craters of all sizes.

Larger moons in the solar system were expected to evolve further because of the quantity of radioactive elements they contained. The radioactive decay would have released enormous heat inside the body. As the size of a sphere increases, the surface area and the volume also increase, but the surface area—which sheds heat—increases only by the square of the radius, whereas the volume—which determines how much radioactive material there is—increases by the cube of the radius. Thus, if two objects are formed from the same material, the larger one should have a higher internal temperature because it cannot shed its heat as well. The result is that a planet or large moon might be expected to have melted soon after it formed, with its heavier components sinking toward the core and its lighter materials rising toward the surface. This process is called *differentiation*. External signs of differentiation can often be found in volcanic activity, which has covered part of the surface of the object with relatively smooth material that is obviously younger because it has flooded and erased many of the craters. Differentiation, however, should not have occurred in smaller bodies because the internal heat would not have been great enough.

Radioactive heating steadily subsides over millions of years as the radioactive elements inside a body decay into nonradioactive ones. Unless a moon is of extraordinary size so that its gravity can hold an atmosphere to cause erosion or unless the moon is positioned so that it experiences extensive tidal strain from its planet and fellow moons, the evolution of that satellite is almost at an end. The only changes come from occasional comet impacts[14] and from high-speed particles and ultraviolet energy from the Sun. Impacts create new craters and bury old ones. The solar wind and radiation create chemical changes in some surface materials and generally tend to darken their color.

With modest sizes, between 300 and 1,000 miles (500 to 1,600 kilometers), and in the eternal cold nearly 2 billion miles from the Sun, the moons of Uranus should have been—and were expected to be—cratered heaps that had "died" as soon as they had formed and with only a pocked faceful of scars to show for their passive existence.

But the moons of Uranus forgot to read that script and act accordingly.

Oberon

Oberon, the outermost moon, came close to fitting expectations. Here, as anticipated, were craters of all sizes, some over 60 miles (100 kilometers) in diameter, in suitable proportions to indicate that Oberon had withstood the initial cratering by large and small objects as the solar system formed about 4.6 billion years ago and had endured the continued cratering by smaller objects as the moons completed their formation around the planet.

At the horizon on Oberon was a mountain, probably the central peak of an impact crater perhaps 200 miles (300 kilometers) across, that rose

Oberon, the outermost moon of Uranus, seen in high resolution by *Voyager 2*.
NASA/Jet Propulsion Laboratory

to an altitude of at least 12 miles (20 kilometers)—a monstrous protrusion more than twice the height of Mount Everest on a world one-eighth the Earth's diameter. The fact that large craters were still visible everywhere on the surface indicated that nothing very much had happened on Oberon for the past 4 billion years.

But Oberon bore a warning that Uranian moons would not conform to expectations. On the floors of a few craters was smooth dark material, indicating that something had welled up from the interior, seeped onto the surface, and solidified. Such a comparatively small moon should not have experienced an interior melt.

Titania

Ostensibly, Titania was the twin of Oberon, almost identical in size, density, color, and reflectivity. Titania was a little closer to Uranus and a tiny bit bigger—1,000 miles (1,610 kilometers) in diameter to Oberon's 960 miles (1,550 kilometers). Neither was half the size of the Earth's Moon. But amid a welter of small craters on Titania were only two or three traces of large craters from the early days of the solar system. And yet Titania must have suffered at least as many large impacts as Oberon. Something had erased those large craters from the face of Titania. How had it happened to Titania but not Oberon?

And that was just the beginning of the contrast. The terrain of Titania was crisscrossed by giant cracks, some 1,000 miles (1,600 kilometers)

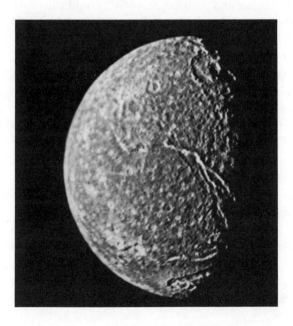

Titania, the largest moon of Uranus, seen in high resolution by *Voyager 2*. The giant cracks, as much as 3 miles deep and 1,000 miles long, expose light-colored material. *NASA/Jet Propulsion Laboratory*

long, 30 miles (50 kilometers) across, and as much as 3 miles (5 kilometers) deep. (For comparison, the Grand Canyon on Earth is one mile deep.) These cracks exposed lighter-colored material.

From the way features overlaid features, geologists could piece together the evolution of Titania. Debris left over as the planets and moons formed pummeled Oberon and Titania, but Titania was a bleeder. Soft ice oozed forth, flooding some features. The crust softened and other features sagged. The heat that caused this partial melting subsided and Titania froze from the outside in. In great abundance among the chemicals that rose toward the surface was water. Water is a most peculiar substance: As it freezes, it expands rather than contracts. The freezing water expanded and stretched the crust until it ruptured in a network of faults, forming steep scarps and deep grabens. Throughout this period of crustal fracturing, additional fluids probably poured forth, covering rough terrain and then freezing smooth and hard. But what was the source of the heat that caused the interior of Titania and perhaps, at least to some degree, Oberon to melt?

Voyager 2 provided clues by revealing the densities of the moons. Pictures from the space probe, taken at known distances, allowed the dimensions of the major moons to be refined to greater accuracy. Tracking *Voyager 2* by its radio transmissions also allowed celestial dynamicists to measure how much the spacecraft had been deflected by the gravity of Oberon and Titania, so that the masses of these two largest moons could be calculated.

With a knowledge of the mass and volume, scientists could calculate the satellites' densities—between 1.4 and 1.7 times as great as water. Such densities are far below those of rocky worlds such as the Earth and Moon. But they are higher than those of the comparably sized icy moons of Saturn, so Oberon and Titania must have proportionally more rock in them. They are perhaps about half rock and half ice. Although the masses and hence the densities of the other Uranian moons could not be measured so accurately, it is likely that they have much the same composition. The greater the abundance of rock, the more radioactive elements would have been incorporated into each moon as it formed, and thus the greater the likelihood that these moons experienced substantial internal heating due to radioactivity.

The present surface temperatures of the satellites of Uranus are approximately $-315°F$ ($80°K$). The satellites of Uranus are actually warmer than the cloud tops of their planet because they have darker surfaces, which absorb more energy from the Sun. To raise water to its melting point would require a vast amount of energy. Under ordinary conditions on Earth, water melts at $32°F$ ($0°C$; $273°K$)—almost $350°F$ above the temperature of the Uranian satellites. Even tidal strain together with radioactive decay cannot account for that much of a temperature difference in such small bodies that rapidly lose so much heat to space. But maybe it was not standard water ice that melted. Perhaps it was methane clathrate—water ice with occasional molecules of methane trapped in the ice crystal lattice. Such a compound melts at a significantly lower temperature, as do other similar compounds such as carbon monoxide clathrate and ammonia hydrate.[15]

Umbriel

Oberon is rather evenly cratered, without much variety to its terrain. Titania has craters but also vast systems of cracks and ice floes—a world of extensive tectonic activity in its early years. One might have expected Umbriel, still closer to Uranus, to continue this progression of increasing geological activity. But the characters of Shakespeare and Pope are capricious, and their satellite namesakes are no different. They defy orderly expectations.

Instead of showing global geologic activity, Umbriel turned out to be almost totally bland. It is the darkest of the major moons, with little variation in its color. It displays no craters with bright rays like those on Oberon and Titania, presumably formed by the relatively recent impact of comets that melted subsurface water ice and splashed the momentarily liquid water across the surface. There still is no generally favored explanation for why Umbriel is so uniformly dark and has so little evidence of tectonic activity while the moons on either side of it exhibit great geologic diversity.

Umbriel, darkest of the Uranian moons, shows craters but little geologic activity—except for two bright features near the top of this high-resolution *Voyager 2* image. Light-colored material is visible at the bottom of a crater 50 miles in diameter and on the slope of the central peak in a nearby crater. *NASA/Jet Propulsion Laboratory*

Yet Umbriel is not completely without contrast. There are two bright features visible. One is on the slope of the central peak of a crater. The other is a ring 50 miles (80 kilometers) in diameter that covers the floor of a different crater. From a distance this circular bright spot gives Umbriel the face of an organ grinder's monkey with its little hat on askew. This brighter material must have come from below the surface.

Since little has disturbed the primordial craters of Umbriel, perhaps the dark surface is original as well. With a diameter of only 740 miles (1,190 kilometers), Umbriel is smaller than Oberon and Titania and therefore should not have experienced internal melting and differentiation except under very unusual circumstances. Perhaps, then, the surface we see is the original crust: very dark carbon-rich rock mixed rather evenly with lighter-colored ice.

Alternatively, it may be that the deep gray surface of Umbriel is just a fairly recent coating from an unknown source that blanketed Umbriel—and Umbriel only. Titania and Ariel, the moons on either side of Umbriel, have no such dark cover.

Whether the darkness of Umbriel's surface is primordial or more recent, the bright ring on the crater floor may be evidence that a relatively recent large impact could temporarily melt enough subsurface ice to allow a significant outpouring of lighter-colored materials to the surface.

Ariel

Ariel is Umbriel's twin in size but in no other way. Umbriel is the darkest of the major Uranian moons, reflecting only 19 percent of the light that hits it. Ariel is the lightest of the moons in color, with an albedo of 40 percent. Umbriel retains one of the most heavily cratered—hence oldest—surfaces in the Uranian system. Ariel's surface is one of the least cratered—hence youngest.

So despite its size and position among the satellites of Uranus, Ariel more nearly resembles Titania than Umbriel. Yet the violence done to Ariel's surface is far beyond what Titania experienced.

Virtually all evidence of large craters and much of the evidence for smaller craters has been obliterated on Ariel. Its surface has been remolded more intensively, over a broader area, and through a longer period than Titania's.

Titania shows extensive fault systems. Ariel displays a global network of faults with rift valleys in some places 10 to 20 miles (15 to 30 kilometers) deep. Some areas on Titania have been smoothed by volcanic floes; on Ariel most of the terrain has been resurfaced by a volcanic process. But the "lava" is not molten rock like volcanoes on Earth produce. Instead it is probably a glacierlike mixture of ice and rock only moderately

Ariel has fault valleys up to 20 miles deep that extend most of the way around the moon, as revealed by *Voyager 2*. *NASA/Jet Propulsion Laboratory*

warmer than the surface of Ariel. This viscous material probably flowed like a glacier too. In places, it rode up over a crater wall and stopped, forming steep scarps more than half a mile high.

As on Titania, but to a much greater extent, upwelling water (probably in clathrate form) froze and expanded, shattering the crust into systems of spectacular deep, narrow valleys. The landscape of Ariel has the look of devastation.

Where did the heat for this volcanism come from? No one knows for certain. Some of the heat came from the decay of radioactive elements; perhaps part from the gravitational compression of the moon as it formed; maybe some from the impact of debris upon its surface; and some, perhaps most, from tidal strain exerted by its neighboring moon Umbriel if the two were temporarily locked in resonant orbits in the past.

Whatever its cause, the source of this volcanic resurfacing faded away at least 3 billion years ago, leaving the gardening of the terrain to the impact of wayward comets.

Miranda

As surprising as Titania was in contrast to Oberon and as Ariel was in contrast to Umbriel, Miranda was as astonishing a single object as *Voyager* had seen in all its travels. A "brave new world," in the words of Shakespeare's Miranda. Here, helter-skelter, side by side, was almost every kind of exotic terrain in the solar system: sinuous valleys, reminiscent of those on Mars, carved by flowing water; grooves in the landscape like the faults on Jupiter's moon Ganymede; cratered highlands similar to the ancient terrain on Earth's Moon; and much more.

Miranda, innermost of the large moons of Uranus. The bright V-shaped feature (the "chevron") lies close to Miranda's south pole. The south polar regions of Uranus and its satellites were pointed sunward when *Voyager 2* flew past. The chevron lies within one of the three peculiar ovoids that *Voyager 2* discovered on Miranda.

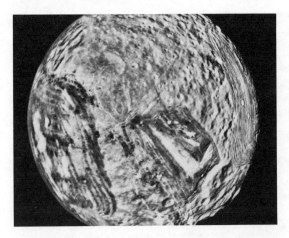

To the left is the ridged and grooved terrain of a much larger ovoid. A third ovoid extends over the horizon to the right.
NASA/Jet Propulsion Laboratory; mosaic by Patricia M. Bridges and Jay L. Inge, U.S. Geological Survey

Voyager 2 flew by Miranda at a distance of 17,500 miles (29,000 kilometers), the closest the spacecraft had been to any of the 51 bodies it had encountered in its eight-and-a-half-year sojourn. Because of this close encounter, the *Voyager 2* mission team could take pictures of Miranda that showed detail down to 2,000 feet (less than one kilometer) across—about the size of a large football stadium. Finer detail could be seen on Miranda than at any of the *Voyagers'* other exploratory sites. And Miranda rewarded the attention amply.

There on Miranda, covering most of the visible surface, was the always-expected mix of large and small craters—indicative of the beating the little moon took after it formed. That much was as it should have been for a small moon. And that was about all there should have been.

But also there, absurdly superimposed on the rolling cratered plains, were three enormous ovoid regions, extending 125 to 200 miles (200 to 300 kilometers) across Miranda—half or more of its diameter. Within these ovoids was light and dark material arranged in ridges, grooves, and scarps, intersecting one another chaotically. There were fewer craters within the ovoids, so the ovoids were apparently a slightly more recent and more localized phenomenon than the earliest cratering. At first glance, two of the ovoids looked like a giant farmer had gone berserk plowing his fields. The third and smallest ovoid, located near the moon's south pole, exhibited a single bright V-shaped feature (dubbed the "chevron"), as if the crazed farmer had laid out an immense boomerang for defense.[16]

Yet the weirdness of Miranda did not end there. The ovoids and the rolling cratered plains were themselves cut by huge systems of fractures that circled the entire moon, creating fault valleys with steep terraced walls. One could stand at the edge of one of these cliffs and gaze down

Miranda is a medley of all the weird terrain in the solar system. Fractures up to 12 miles deep cut through the ovoids and encircle the moon.
NASA/Jet Propulsion Laboratory

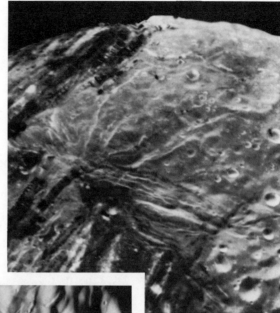

Valley walls on Miranda catch the sunlight while night blankets the surrounding area, revealing how high and steep the cliffs are. Miranda is only 300 miles in diameter but has canyons more than 10 times deeper than the Grand Canyon on Earth.
NASA/Jet Propulsion Laboratory

into a canyon 6 to 12 miles (10 to 20 kilometers) deep—roughly ten times the depth of the Grand Canyon on Earth and up to three times the depth of the Mariner Valley chasm on Mars. If an astronaut in the weak gravity of Miranda jumped off a 12-mile-high cliff, he would fall for ten minutes before he hit the valley floor. It would not be a happy landing. At first he would fall slowly, but then gradually he would fall faster and faster until he reached the bottom of the chasm at a speed of 120 miles per hour (200 kilometers per hour).[17]

And just to round out this smorgasbord of a surface, Miranda displayed some evidence of floes from eruptions long ago, although not as extensively as Ariel.

"Isn't it wonderful?" said one scientist admiring Miranda. Said another: "It looks like a satellite designed by a committee."[18]

About all the geologists could agree upon was that Miranda was one of the strangest worlds yet seen. Its face was more what might have been expected from a planet or moon ten times its size. With a diameter of only 300 miles (484 kilometers), Miranda should scarcely have had enough gravity to pull its mass into a spherical shape. Yet Miranda had not been passive after its formation. Here, in the ovoids, the fractures, and the flows, was evidence that it had once had considerable internal heat, enough to let its interior at least partially differentiate.

One explanation for the erratic surface of Miranda was that the satellite was initially formed with a not completely uniform mixture of rock and ice throughout. Then internal heat allowed the start of differentiation. The heavier materials began to sink toward the center and the lighter materials began to float toward the surface, but three to four billion years ago the internal heat was fading and the moon was freezing. The lighter and brighter ice buoyed up to the surface in only a few places to form the ovoids. It was a case of arrested development. Had differentiation continued, the bright ovoid material would have covered the whole moon and smoothed all the terrain.

It is most strange indeed that geologic activity ever began on Miranda or Ariel, considering their small size. Ariel is only one-third the diameter of Earth's Moon, Miranda only one-seventh. They are too small in volume to have had enough radioactive elements to allow for even partial internal melting if their composition resembles the other bodies in the solar system. Yet the evidence of internal heat is undeniable.

Where did it come from? The most likely explanation is that much of the heating was provided by tidal strain induced by Ariel in a resonant orbit with Miranda long ago. If Miranda caught up with and passed Ariel at the same point in its orbit at regular intervals for an extended period of time, the pull on Miranda from Uranus in one direction and from Ariel in the opposite direction would have the effect of flexing the tidal budge of Miranda, creating significant frictional heat inside the body. Tidal heating is the best explanation for the volcanoes on Jupiter's Io, the cracks in the ice of Jupiter's Europa, and the floes that truncated craters on Saturn's Enceladus.

In the present age Miranda, Ariel, and Umbriel are not in tidal resonance, but they may have been at some period in the not-too-distant past. If so, the Miranda-Ariel connection may have given Miranda its haphazard terrain, and the Ariel-Umbriel resonance could have given Ariel its fractured look. Umbriel may in turn have experienced tidal strain imposed by Ariel or perhaps Titania, but Umbriel's greater distance from Uranus may have kept the strain within manageable limits, leaving the satellite with an unstretched face. Thus, despite being the same size, Umbriel and Ariel may have developed differently because they lie at different distances from Uranus.

Ten New Moons

Voyager 2 not only unveiled the five known moons of Uranus; it also discovered ten new ones, tripling the number of documented satellites for Uranus. Yet astronomers had hoped to detect at least 18 new moons.

The reason for such high expectations was the need to explain the nine narrow, sharp-edged rings of Uranus. It was reasonably easy to explain the appearance of Saturn's broad, flat disk of rings. Each particle in the rings is a separate satellite with an orbit all its own. The particles, by the billions, must occasionally collide, thereby altering their orbits. They also knock chips off one another and scatter those fragments onto unique orbits of their own. In this way, a ring of chunks steadily pulverizes itself and widens into a band of debris.

In examining the rings of Saturn over the centuries, astronomers on

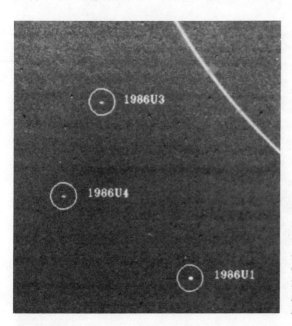

Voyager 2 discovered ten new small moons at Uranus. Three of them are circled here, lying outside the outermost ring of Uranus.
NASA/Jet Propulsion Laboratory

Earth also noticed a series of apparent gaps of a few thousand down to a few hundred miles where there were no ring particles or at least relatively few. But this too was explainable if one relied on the moons of Saturn. Each moon has its orbit. Each ring particle is a tiny moonlet with an orbit too. Imagine a ring particle named Sam with an orbital period of 11 hours. Saturn's moon Mimas, in a higher orbit, has a period of 22 hours, twice that of Sam. So every time Sam goes halfway around Saturn, he catches up with and passes Mimas. Each time he passes Mimas, Sam experiences a gravitational pull outward, away from Saturn. He is pulled into a slightly higher orbit. These tugs occur twice during each revolution at the same two points on opposite sides of Sam's orbit, so the orbit remains nearly circular but is now farther from Saturn. In a higher orbit, Sam revolves more slowly. He no longer catches up with Mimas at the same points along his orbit. The gravity of Mimas still tends to raise Sam's orbit, but there is no balancing tug on the opposite side of the orbit or at symmetrical points along the orbit. The result is that Mimas' pull now tends only to make Sam's new orbit more elliptical, which causes Sam to collide with other particles. These collisions tend to return Sam to a more circular orbit.

Thus Saturn's major inner moons, through their orbital resonance with particles in Saturn's rings, tend to clear out or reduce the number of particles at certain distances in the ring system.

Ring theory was in good shape until the rings of Uranus were detected. Astronomers could understand broad, flat ring systems. They could understand narrow gaps between rings. But here, revealed in the 1977

The complete ring system of Uranus. Nine thin, dark rings had been discovered from Earth. *Voyager 2* found two more.
NASA/Jet Propulsion Laboratory

star occultation by Uranus, were extremely narrow rings with broad gaps. Astronomers tried to explain what they detected by satellite resonance. It didn't work. The moons of Uranus were too small, the gaps were too large, and the rings were too finely "sculpted."

Theoreticians went to work. Peter Goldreich and Scott Tremaine produced an intriguing answer.[19] The particles in each narrow ring were herded and kept in position by at least one and probably two moons acting as shepherds. Picture Epsilon, the brightest and outermost ring of Uranus. Left alone, collisions would be constantly slowing some particles in the rings so that they would fall to lower orbits. Other particles, however, would be gaining speed by collision and would rise to higher orbits. Inevitable and continuous collisions among the particles in the Epsilon Ring or any ring would cause it to spread out—and rather quickly too.

But now imagine a small moon orbiting Uranus just inside this ring and a second small moon just outside. The closer an object is to the body it orbits, the faster it travels. So the inner shepherd would be orbiting faster than the ring particles. Every time it passes a ring particle, its gravity gives that particle a little tug—a little extra speed—and that boost causes the particle to rise to a higher orbit. The inner shepherd uses its gravitational energy to prevent scattered particles from falling inward from the ring.

The outer shepherd moon, however, is circling more slowly than the ring particles. Each time a ring particle passes that moon, the moon's gravity tends to reduce its speed and it begins to fall to a lower orbit. In this way, particles with excess speed that are rising out of the ring are slowed and dropped back to the ring once more.

So the outer shepherd shears the outer edge of the ring and the inner shepherd grooms the inner edge. In this way, a narrow ring can last an extended period of time.

But the explanation is not yet complete. The conservation of angular momentum governs both the rings and the shepherd moons. Each time the inner shepherd adds velocity to a ring particle, the moon loses a minute amount of energy and falls to a minutely lower orbit. Each time the outer shepherd retards the velocity of a ring particle, it gains that same amount of energy and rises to a minutely higher orbit. How then do the shepherd moons stay in place to continue their ring-preservation duties? The answer appears to be resonance with the larger, outer moons.

This was the explanation that Goldreich and Tremaine offered for the thin rings of Uranus. Since the required shepherd moons would be too small to detect from Earth, it would be up to *Voyager 2* to attempt to find these moons when it arrived at Uranus in 1986.

Two shepherd moons (marked), one on each side of the Uranus' largest and outermost (Epsilon) ring, provide the gravitational effects that confine the particles and keep the ring narrow. Theorists predicted their existence; *Voyager 2* found them for the Epsilon Ring but was not able to see shepherds for the other Uranian moons. Perhaps they are too small.

NASA/Jet Propulsion Laboratory

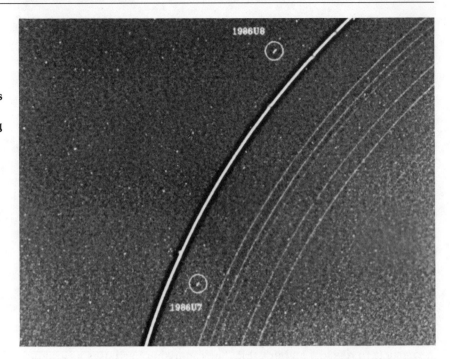

But the confirmation of shepherd moons did not wait that long. NASA's *Pioneer 11*, the first spacecraft to fly by Saturn, discovered in 1979 that Saturn had a faint, narrow ring (now known as the F Ring). When *Voyager 1* arrived at Saturn in 1980, it found that the F Ring had a pair of shepherd moons.

So when *Voyager 2* reached Uranus in 1986, astronomers expected to see two shepherd moons for each of the nine known rings of Uranus.

A pair of shepherd moons—about 25 to 30 miles (40 to 50 kilometers) in diameter—was discovered for the Epsilon Ring. Eight other small satellites were discovered, closer to Uranus than the major moons but all beyond the rings. They were not shepherds, and their gravities were too small to significantly affect the rings by resonance.

Where were the expected shepherds? At *Voyager 2*'s closest approach to the rings, its cameras could detect moons larger than about 12 miles (20 kilometers) in diameter, but the shepherds required by the rings of Uranus could be as small as about 10 miles (15 kilometers) in diameter. There may well be small undiscovered objects lurking in orbit around Uranus, especially pairs of shepherd moons dogging each ring, but except for the shepherds of the Epsilon Ring, careful analysis of *Voyager 2*'s full ring-survey images has not revealed them.

So Uranus, with 15 known moons, may yet have twice that number of satellites. But we probably won't know for certain until the successor to *Voyager* reaches that world of many worlds with improved imaging equipment. Unfortunately the United States has no such mission in preparation. There may be no more close-up spacecraft explorations of Uranus or Neptune in our lifetimes.

The first of the ten new moons of Uranus that *Voyager 2* discovered was detected on December 31, 1985, while *Voyager 2* was still three and a half weeks from Uranus. Pending official confirmation, the moon was unpoetically labeled 1985U1. But the mission team dubbed it Puck, raiding Shakespeare's *A Midsummer Night's Dream* once more.

Because of Puck's early detection, its substantial size, and its position closer to Uranus than the five major moons, the picture-taking schedule of *Voyager 2* was modifed to allow for one close-up image of Puck by sacrificing one picture to be taken of Miranda.

The close-up of Puck showed a dark, uneven surface about 105 miles (170 kilometers) across—a little more than one-third the diameter of Miranda. Puck was almost spherical in shape and exhibited some muted large craters. Perhaps the outlines of these comparatively large craters had been eroded over the eons by the impacts of millions of tiny particles whose craters were not visible from a distance of 320,000 miles (515,000 kilometers).

As nearly as *Voyager* could tell, the nine other new moons that it discovered were smaller versions of Puck. They had diameters ranging from about 50 miles (80 kilometers) down to 25 miles (40 kilometers). All lay closer to the planet than the major moons and Puck, and all were very dark in color, with a reflectivity less than the Earth's Moon.

The ten new moons, all much darker than the major moons, are about the same charcoal black color of the rings. The rings and these inner moonlets may therefore have the same composition and the same origin.

Some *Voyager* scientists speculate that the moonlets and ring particles are covered with methane ice whose original light color has changed chemically over the ages through exposure to high-energy protons trapped in the magnetic field of Uranus. This energy breaks down methane into carbon and hydrogen and polymerizes the carbon into long-chain hydrocarbons, black in color, dubbed "star tar" by Carl Sagan, one of the proponents of this theory.

Other *Voyager* scientists explain the dark color of the small satellites and rings as carbon compounds present from the origin of the solar system, rather than recently blackened. It is the same mixture of ice and dark carbonaceous materials that formed Uranus, says planetary scientist Robert Hamilton Brown. He calls this debate between interpretations "tar wars."

Regardless of its origin, the dark color of the moons and rings was one of *Voyager 2*'s most important findings at Uranus. Earlier in 1986, Soviet and European space probes that flew by Halley's Comet discovered that its nucleus was also extremely dark. This blackish hue suggests abundant carbon or carbon compounds, vital for life. Thus, satellites in the outer solar system and their cometary antecedents appear to be emphasizing that our solar system (and others?) formed with the right chemical elements for life to begin wherever conditions would permit.

Busted Moons and Ring Fodder

Eleven dingy rings. There isn't much to them. Prior to the *Voyager 2* flyby, nine rings of Uranus had been known since 1977 when they revealed their existence by taking turns momentarily blocking the light of a star. All were thin in dimensions and material. The Epsilon Ring, the outermost and most substantial of the eleven, varies in width from 14 to 58 miles (22 to 93 kilometers). The other rings have widths of from 0.5 to 7 miles (1 to 12 kilometers).[20] If Uranus were shrunk to the size of a golf ball, the rings would be the width of strands in a spider's web.

The rings of Uranus may be sparse and black as coal, but they are relatively free of dust. In fact, observations by *Voyager* at Uranus showed comparatively few particles smaller than beach balls or chairs. By contrast, Jupiter's ring system is by mass about 50 percent dust particles measuring about one-ten-thousandth of an inch (a few microns) in size—about the dimensions of the tiny solid grains in smoke that make it opaque. Saturn's rings have 10 times less dust than Jupiter's. And the rings of Uranus have about 100 times less dust than those of Saturn.

Voyager 2 could measure the dustiness of the rings as it passed Uranus by looking back so that the sunlight struck one side of the rings while *Voyager* observed from the other. Under this circumstance, good-sized ring chunks (a foot, a yard, a mile in size) are hard to see, but tiny solid flecks of ice the size of dust are conspicuous because they tend to scatter light forward. The effect is like what a moviegoer sees in a darkened theater as the film is projected. Looking down the projector's beam of light toward the screen, one usually sees only reflected brightness from the screen. But from the front or sides of the theater, the light's path from the projector toward the screen is obvious as it illuminates dust in the air.

So the dust in the rings of Uranus—what little there was—showed up when *Voyager 2* put the rings of Uranus between itself and the Sun. This scarcity of dust in the Uranian rings was startling. It indicated that the rings of Uranus must be evolving continuously and rapidly. The particles in any ring system must be colliding, and thereby grinding themselves down. A ring system in equilibrium must contain a sizable amount of

Past Uranus, *Voyager 2* looked back through the rings toward the Sun. The image revealed dust particles, although far fewer than in the rings of Jupiter and Saturn. Since ring particles continue to form dust by collisions, the dust must be falling into the atmosphere of Uranus. The ring system of Uranus is eroding rapidly.
NASA/Jet Propulsion Laboratory

fine dust. Jupiter's small dusty ring may be the remnant of collisions and disintegrations that have worn the chunks down to fine powder. Saturn's icy rings contain a balance of large and small fragments. At Uranus the large particles are present, but the small particles are missing. The rings of Uranus are not in collisional equilibrium. The large fragments are colliding and making dust. But something is removing dust from the rings of Uranus almost as fast as it is formed.

Voyager 2 framed the problem. *Voyager 2* found the answer. Its ultraviolet spectrometer showed that the hydrogen atmosphere of Uranus extends in a very tenuous form much farther from the planet than expected, all the way out through the rings. The cause of this bloated atmosphere was assigned to the phenomenon that creates dayglow. In that process, hydrogen atoms are slowly but steadily escaping from Uranus, and as they do they provide sufficient density to exert a significant drag on the dust particles in the rings. The dust particles are steadily slowed and fall to ever-lower orbits until they disappear from the ring system and vanish into the shroud of gases in the atmosphere of Uranus.

The entire Uranian ring system is rapidly eroding.

Far from Uranus, where Oberon lies, there is little evidence that anything much has happened geologically since the intense meteoroid bombardment of the solar system's early days. Closer to Uranus, Ariel and Miranda are smaller worlds with extended geologic activity written on their faces. Still closer to Uranus are the ten new moons, all tiny by comparison and, to judge by the largest one, all thoroughly battered. Still closer than the moons is the Epsilon Ring, the densest, widest, and brightest of the rings, with two of the newly found ten moons serving as shepherds. And still closer are the remaining ten rings, composed of chunky debris.

But why should the moons of Uranus show evidence of greater, longer, and more disruptive geologic activity the closer to the planet they lie?

The internal heat from the decay of radioactive elements and gravitational compression necessary to melt the interior and remold the surface ought to be greatest in the largest moons. Yet Oberon and Titania are the two outermost and do not show the greatest geologic activity. Instead, the most profound geologic upheaval has occurred on Ariel and Miranda, the two innermost and smallest of the five major moons. Tidal strain induced by Uranus and neighboring moons in resonant orbits may well have been a principal cause. But another factor may have contributed as well.

Miranda and Ariel were battered more severely by impacts than the others, says the hypothesis, for the very reason that they lie closer to Uranus. The debris orbiting the Sun left over from planet accretion was diverted in its course by the enormous mass of Uranus so that it passed closer to the planet, creating a region around Uranus where the density of flying junk was higher and the chances of a moon getting hit were greater. Uranus gravitationally focused a hail of debris on itself and its satellites, especially the innermost ones.

Judging from the ancient face of Oberon and by invoking the concept of gravitational focusing by which Uranus hurled planetesimals upon its huddled satellites, the *Voyager 2* imaging team estimated that, compared to Oberon, Ariel would have received 5 times as many impact craters and Miranda 14 times as many. Uranus was guilty of child abuse.

But how does anyone know about the frequency of cratering events in the Uranian system? The process begins with an estimation of the number of comets passing nearby.

Comets are essentially dirty snowballs, made mostly of water ice with flecks of dust mixed in. The solid nucleus of a typical comet is about 5 to 10 miles (8 to 15 kilometers) in diameter. Based on the orbits of comets observed from Earth, Dutch astronomer Jan H. Oort established in 1950 that all known comets were members of our solar system, held captive by the gravity of the Sun, and not itinerants from outer space just passing through our solar system. Therefore, said Oort, to account for the steady trickle of comets that pass close enough to the Sun for us to see, our planetary system must be surrounded by a sphere of comets, starting beyond Neptune and Pluto and extending halfway to the nearest star—out to a distance of about two light-years.[21] In this realm, called the Oort Cloud, there must be, according to recent estimates, trillions of comets remaining today, even after 4.6 billion years of attrition.

The comets in the Oort Cloud lie at such distances that we cannot detect them from Earth. They travel in such remote orbits that we would never see any of them except that, every once in a very long while, the Sun, dragging its family of planets and satellites along, passes in the vicinity of another star. When that happens, depending on which directions the star and comets are going, the gravity of the passing star accelerates millions of comets and they escape from our solar system forever. But the passing star also decelerates millions of other comets, and they begin a long fall that will carry them over a period of many centuries closer to the Sun on new orbits. Most of those falling millions never get as close to the Sun as is Pluto or Neptune. But the few that do enter the planetary realm of the solar system and penetrate deeply to within Mars' or Jupiter's distance from the Sun show us by their orbits that they are gravitational members of our Sun's family and have come our way from the outer limits of our Sun's gravitational control.

Based on the comets we detect and on statistical probability, for every comet whose orbit starts outbound again at about the distance of Jupiter, there should be twice as many comets that venture no closer to the Sun than Saturn and four times as many that approach no closer than Uranus.

Comets are small in size, so their paths are easily disturbed by the gravity of the giant planets. These perturbations hurl some comets out of the solar system. Other comets will be slowed so that they cannot return to the Oort Cloud. Instead they are trapped, at least temporarily, on elliptical orbits relatively close to the Sun. Before, they had orbital periods of thousands or millions of years. They were long-period comets. Now they are short-period comets, with periods of 200 years or less.

Because the gravity of Uranus gets a crack at every inbound long-period comet before Saturn or Jupiter do, many more comets are expected to show orbital evidence of a gravitational encounter with Uranus than with Saturn or Jupiter despite their far greater masses. Some comets have been diverted so that they never venture much farther from the Sun than Jupiter. They are said to be the Jupiter family of comets. So too there are families of comets with aphelia at about the distances of Saturn and Uranus. But because far more comets reach inward only as far as Uranus rather than Saturn or Jupiter and because Uranus gets first gravitational crack at all that do, some astronomers estimate that Uranus has a harem of short-period comets 100 times more numerous than Saturn and 600 to 700 times more numerous than Jupiter.

Uranus finds itself constantly in the midst of a snowball fight with iceballs the size of Earth mountains or larger. There are probably a million short-period comets that frequently pass near Uranus.[22] Over time, Uranus and its satellites stand a good chance of getting hit. Uranus faces a larger army of assassins than Jupiter or Saturn, but it has a larger area of space in which to hide and less gravity to draw trouble its way. Jupiter and Saturn may face fewer prowling assassins, but they have smaller orbits along which to scurry to escape a hit and more gravity to focus comet attention toward them. The result is that the number of impacts in the Jovian, Saturnian, and Uranian systems evens out. A square mile on Uranus and its satellites may be expected to be blasted with about the same frequency as a square mile on Jupiter, Saturn, and their satellites.

The frequency of comet cratering can be estimated well enough, comet specialists believe, so that when estimates of comet damage are compared with the observed damage to the moons of the giant planets, conclusions may be drawn. Oberon and Umbriel show too many craters, especially large ones, to have been cratered only by comets. They are survivors of at least most of the earliest formative events at Uranus.

The ten newly found inner satellites of Uranus might well be fragments from moons that shattered about 3 billion years ago. The rings of Uranus are most likely fragments from the ongoing demolition of its satellites. If all the material in the Uranian rings could be scooped up and packed into one large rocky iceball, it would form a body only 18 miles (30 kilometers) in diameter, smaller than any moon yet discovered at Uranus. By comparison, the particles in Saturn's ring system would form a body 180 miles (300 kilometers) in diameter.[23] The rings of Saturn contain a thousand times more mass than the rings of Uranus.

The Price-to-Learning Ratio

Long before *Voyager 2*'s data could be fully analyzed and its discoveries thoroughly appreciated, Uranus was a starlike dot far behind the spacecraft, which was already preparing for its next and final port of call.

From launch through the encounter with Uranus, *Voyager 2* had cost each American $2. It was a remarkable return on an investment.

CHAPTER 10

Appointment with Neptune

"The real heroes are the engineers at JPL who built the wonderful spacecraft that have lasted so long and performed so well."
Theoretician Scott Tremaine
(1988)

"For a spacecraft designed and built back in the days when hand-held calculators were first being marketed, the Voyagers *have been remarkably responsive to the science and engineering demands placed on them."*
Ellis D. Miner, Assistant Project Scientist for *Voyager*
(1986)

Because of its great distance, it has been hard to pry secrets from Neptune. It is about the same size as Uranus but more than half again as far away, so it is about ten times fainter and offers even less of a disk upon which detail might be discerned. One large moon, Triton, was found soon after Neptune was discovered, but a century passed before a second moon, Nereid, was found in 1949.

Neptune moves so slowly in its distant orbit that, since its discovery in 1846, it has not yet had time to complete one revolution around the Sun. That event won't occur until 2011.

For *Voyager 2*, Neptune was a staggering challenge because it stood last in a line of ifs. If *Voyager 2* survived, collected data, and achieved the proper gravitational assist at Jupiter; if . . . at Saturn; if . . . at Uranus. Each flyby placed its own special demands on the spacecraft's trajectory, and the demands restricted its flight path at the next planet.

If scientific success at Uranus were to match the *Voyagers'* triumphs at Jupiter and Saturn, the researchers needed the spacecraft to pass close to the planet; penetrate deeply into the planet's suspected magnetic field; sail close by the planet's innermost major moon, Miranda; and then swing behind the planet and its rings so that the scientists could use the radio transmission of *Voyager* passing through the planet's ring system and atmosphere to reveal its structure and composition.

Voyager 2's flight by Uranus was also tightly constrained by the spacecraft's need to use the planet for a gravity assist to reach Neptune. But for *Voyager 2* to get just to the vicinity of Neptune was not enough. Researchers wanted a full-fledged scientific investigation, which meant, of course, flying close to the planet, plunging deep into its magnetosphere, and beaming radio signals through its rings (if there were any) and atmosphere. It meant, as well, passing close to Neptune's largest moon, Triton, and flying behind Triton so that *Voyager*'s radio signals could pass through the moon's atmosphere to indicate its structure and composition.

It was hard enough to meet the objectives at Uranus even if one could neglect the gravitational-assist requirements to divert the craft on to Neptune on a precise billion-mile course. To meet all the objectives at both Uranus and Neptune was a lot to ask of one single flight path.

Yet such a flight path was found. It would carry *Voyager 2* only 2,700 miles (4,400 kilometers) above the cloud tops at Neptune's north pole, then behind the planet, and then on past the moon Triton at an altitude of only 25,000 miles (40,000 kilometers).[1] With utmost skill and care, *Voyager 2* could even avoid colliding with the planet, its partial rings, and its moons. It would be the closest planetary flyby in the *Voyager* program. To make it work, *Voyager 2* at Uranus had to hit a point in space at a point in time with phenomenal accuracy. To miss "corridor center" at Uranus by a mile would create an error of 4,000 miles at Neptune. A small miss at Uranus could be corrected later by using the spacecraft's thrusters to adjust speed and direction, but such a maneuver was limited by the fuel remaining.

The accuracy required of the spacecraft navigators at Uranus to place *Voyager 2* on its unique trajectory for Neptune, said Charles Kohlhase, *Voyager* mission planning manager, was a feat comparable to a golfer sinking a thousand-mile putt.

The celestial golfers holed out. When the mission team determined that *Voyager 2* was headed down the corridor only 6 miles (10 kilometers) off center, they canceled the final midcourse correction.

As *Voyager 2* passed 50,700 miles (81,600 kilometers) above the cloud tops of Uranus on January 24, 1986, the gravity of the planet bent the spacecraft's trajectory by 23 degrees and increased its velocity by almost 4,500 miles per hour (2 kilometers per second). *Voyager 2* was on course to Neptune. A billion miles and three and a half years of travel lay ahead.

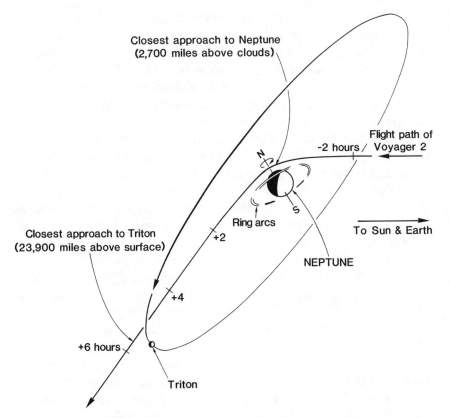

Closest approach to Neptune
(2,700 miles above clouds)

Flight path of
-2 hours / Voyager 2

N

S

Ring arcs

To Sun & Earth

Closest approach to Triton
(23,900 miles above surface)

+2

NEPTUNE

+4

+6 hours

Triton

Voyager 2 at Neptune. On August 24–25, 1989, Voyager 2 will fly outside the ring arcs, over the north pole of Neptune, then on past Triton. Triton is the only large moon in the solar system to travel around its planet in the direction opposite to its planet's spin and revolution around the Sun. Triton's orbit is also inclined 20 degrees to Neptune's equator. After passing Triton, Voyager 2 will be headed out of the solar system in a southerly direction.

The New Challenge for *Voyager 2*

Hiding near the planetary fringe of our solar system, a billion miles (1.6 billion kilometers) farther out than Uranus, Neptune is about four times too dim ever to be seen with unaided eyes. At a distance from Earth of 2.8 billion miles (4.5 billion kilometers), Neptune presents a disk only 2.3 arc seconds across, so discerning features on the planet is like seeing details on a dime a mile away.

Because it is 30 times more distant from the Sun than is Earth, Neptune receives 900 times less sunlight than our planet and 2.5 times less than Uranus. The Sun in the skies of Earth is a disk measuring about half a degree (30 arc minutes) across. From Neptune the Sun is only one arc minute in diameter—the size of a quarter seen across the length of a football field. Yet the Sun in the sky of Neptune is still about a thousand times brighter than the full Moon in our nighttime sky. Even so, the light levels at Neptune are low—the equivalent of twilight about ten minutes after sundown on a clear day on Earth.[2]

Voyager 2 had three successful planetary encounters behind it, but Neptune posed a new set of obstacles. First, the spacecraft would have to survive the three and a half years of flight between Uranus and Neptune. Second, because of Neptune's greater distance from the Sun, *Voyager 2* would have to take pictures at still-lower light levels. Third, because close approaches to Neptune and Triton were scientifically desirable, the speed of the spacecraft would be greater and the hazard of smeared pictures increased. Finally, the increased distance would make radio communication with the spacecraft even more difficult. At a range of 2.8 billion miles (4.5 billion kilometers), instructions radioed to the spacecraft would take four hours and eight minutes to arrive, and the data requested would take another four hours and eight minutes to reach the Earth.

The logistics of Earth's first spacecraft to visit Neptune put exceptional pressure on the one remaining crippled onboard radio receiver and on the fault detection and correction systems of the onboard computers. It also put enormous pressure on the team of spacecraft navigators on Earth who, by using *Voyager 2* pictures of the Neptunian moons against their starry backgrounds, would have to determine the craft's exact location, perhaps modify its speed and angle of approach slightly as remaining fuel would permit, and update the craft about the changed target times and angles for the cameras and other instruments.

In the three and a half years between the Uranus and Neptune encounters, improvements to *Voyager 2* and to the communication system on Earth continued. The objective was to get from *Voyager 2* the same quality pictures and the same rate of data transmission from Neptune as had been achieved at Uranus despite the 57 percent greater distance, which more than doubled the problems of transmission and reception. In the perpetual twilight so far from the Sun, *Voyager*'s television cameras would require longer exposures to form pictures. The longest possible time exposure at Uranus without using the tape recorder had been 15 seconds. For Neptune, due to computer reprogramming, the longest exposure time without using the tape recorder could now be several hours.

Voyager 2 was now even steadier as a camera and instrument platform than it had been at Uranus. By firing the thrusters in bursts of only five thousandths of a second to correct its aim, the spacecraft at Uranus was twice as steady as it had been at Saturn. At Neptune, the thrusters would fire in four-millisecond bursts for a further increase in camera-pointing stability of about 20 percent.

Even so, flying close to Triton's expected landscape of features at high speed would lead to blurred pictures, so the *Voyager* scientists and engineers improved target-motion compensation (panning the spacecraft to follow the action) with nodding image motion compensation. Using small thruster bursts, *Voyager* will turn slightly to take a picture, then close its camera shutter and swing back to point its antenna at Earth so that the picture can be transmitted immediately without being stored on the tape recorder. The craft then nods back to its target to resume picture taking.

But it was not enough just to make improvements to a 12-year-old spacecraft built with early 1970s technology on an extended mission 2 billion miles (3 billion kilometers) and eight years beyond mission requirements. Improvements were made to facilities on Earth as well.

Voyager 2 could not increase its radio-transmitting power, so NASA set about increasing the receiving power on antennae on Earth. At Goldstone, California; Madrid, Spain; and Canberra, Australia, are the tracking stations of NASA's Deep Space Network, each equipped with a steerable radio receiving dish 210 feet (64 meters) in diameter. NASA enlarged all these dishes to 230-foot (70-meter) receivers, increasing signal reception by about 20 percent. Other modifications to reduce background noise provided the big dishes with a total improvement in receptivity of about 50 percent.

In addition, arrangements were made so that the Goldstone, California, dish would be assisted by the Very Large Array radio-telescope facility in Socorro, New Mexico. The VLA is a system of 27 movable dishes each 82 feet (25 meters) in diameter. The Goldstone dish and the Very Large Array receivers would be linked together to more than double their receiving power to catch the less than a billionth of a millionth of a watt reaching the array from *Voyager 2* at Neptune. The Canberra, Australia, dish of NASA's Deep Space Network would also receive help from Australia's 210-foot (64-meter) Parkes Radio Telescope and, when *Voyager 2* passes beyond Neptune and Triton, from the 210-foot (64-meter) dish of Japan's Usuda space-tracking station.

Target Neptune

Uranus and Neptune, say the astronomy textbooks, are pretty much twins. And so they are in size, in mass, in atmosphere, and in color. It is their similarities that provide a basic set of expectations about what *Voyager 2* will find at Neptune.

Uranus is a little larger in diameter, but Neptune has a slightly greater mass. Both have thick atmospheres composed overwhelmingly of hydrogen and helium flavored with small quantities of other gases, including methane. It is the methane that gives both Uranus and Neptune their blue-green color by absorbing the red wavelengths of sunlight.

But there are also differences—and significant ones.

Uranus and Neptune may be similar in size and mass, but because Neptune is smaller in diameter and yet greater in mass, it has a considerably higher density. That density probably enhances Neptune's (not yet detected) magnetic field and internal heat.

**Neptune and Triton
(marked by arrow) as
seen from Earth with
the 120-inch telescope
of the Lick Observatory**
*Lick Observatory
photograph*

Unlike capsized Uranus, Neptune stands relatively upright as it revolves around the Sun. There is some axial tilt—29 degrees—very similar to the Earth's 23.5 degrees. Unlike Uranus, there is never any sunlight beaming down almost vertically on the Neptunian poles.

When it became possible to sample wavelengths beyond the optical light coming from Neptune, astronomers discovered that the total amount of energy radiating from Neptune was greater than the energy it was receiving from the Sun. The excess energy was in the form of heat and radio waves. Jupiter and Saturn emit about twice as much energy as they receive from the Sun due to heat from their interiors. Uranus emits more energy that it receives, but just barely. At Neptune, at least half of the energy flowing from the planet is from internal heat. It is radiating two to two and a half times more energy than it receives. It is not that Neptune is so remarkably warmhearted. It is just that it receives so little energy from the Sun that it is easier to detect the planet's own emission. Neptune's internal heat production may be roughly comparable to that of Uranus. The behavior of *Voyager 2* deep in the gravitational and magnetic fields of Neptune should provide new information about the planet's interior structure, which could clarify the source of this internal energy.

And, perhaps because of its greater interior heat, Neptune, unlike Uranus, shows cloud features that are visible from Earth. It was a struggle to see them, but they are there. Jupiter's clouds are exceptionally distinct and colorful. Saturn's clouds are shrouded by haze but still show considerable variation in hues. Ground-based observers occasionally claimed to see features in the atmosphere of Uranus, but they could not be confirmed by most astronomers. When *Voyager 2* arrived at Uranus, its pictures in visible light showed an essentially featureless atmosphere. Only intensive computer enhancement of contrast showed a very modest array of details undiscernible by the unaided eye from the rest of the atmospheric background.

Based on the progression toward weaker atmospheric detail from Jupiter to Saturn to Uranus, one might have expected no cloud details on the colder Neptune. But Neptune didn't care about expectations.

The discovery of cloud details on Neptune was made possible by a new imaging technology called the charge-coupled device (CCD). Soon after its invention in 1969, the CCD began to outperform photographic film, which had been the picture-taking backbone of astronomy for a hundred years. CCDs are small silicon chips divided into many rows and columns. Photons striking each tiny sensitive spot trigger a buildup of electrical charge in proportion to the number of photons received. By measuring the charge at each site, a picture can be constructed. In this way CCDs serve as the heart of supersensitive television cameras. Using CCDs with Earth-based telescopes, astronomers detected on Neptune bright high-altitude clouds that form and vanish—and provide a means of estimating the rotation of the atmosphere of Neptune at different latitudes.

As on Uranus, the length of a day on Neptune had been hard to measure and was highly uncertain. As a gas giant planet, Neptune was expected to rotate faster than rocky midgets like Earth. As a small gas giant like Uranus, Neptune was expected to rotate slower than the larger gas giants, Jupiter and Saturn. Neptune showed a modest equatorial bulge, which suggested a rotational period as fast as 15 hours.

The extent of the equatorial bulge on Neptune was determined from the planet's occasional occultations of stars. Those same occultations allowed the diameter of Neptune to be measured with respectable accuracy: 30,800 miles (49,600 kilometers).

The determination of Neptune's rotation by spectroscopy and photometry had left considerable uncertainties. Spectroscopy measured how fast Neptune's western limb was rotating toward the Earth and how fast the eastern limb was turning away. Together with the exact diameter of Neptune, these speeds could furnish a reasonable calculation of the rotation of Neptune (or at least its atmosphere).

Photometry measured variations in the light reflected from Neptune so that if it had a long-lived atmospheric feature like Jupiter's Great Red Spot, its appearance and disappearance as the planet turned could reveal

This photograph of Neptune, taken by Heidi B. Hammel with the University of Hawaii's 88-inch telescope on Maura Kea in 1986, is the best ground-based image of Neptune yet obtained. It shows clouds (bright spots) in Neptune's

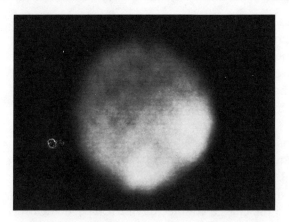

southern hemisphere. From these features, Hammel was able to determine the rotation period for Neptune (or at least the atmosphere at that latitude) to be 17 hours.
Courtesy of Heidi B. Hammel, Mauna Kea Observatory

Neptune's rotational period. One photometric measurement suggested that Neptune rotated in 17 hours 43 minutes.[3]

Finally, in 1979, the first cloud features on Neptune were distinguished. Subsequent observations have refined the rotation period of Neptune (or at least its clouds) to about 17 hours.[4]

The Search for Rings at Neptune

With the discovery of the ring systems of Uranus and Jupiter, most astronomers suspected that Neptune would have rings also. Were they so faint, like those of Jupiter, that evidence of their existence would have to await a visit from *Voyager*, or were they substantial enough, like those of Uranus, to be detected from Earth when they occulted a distant star?

Such occultations, however, are much rarer for Neptune than Uranus for two reasons. First, although the two planets are about the same size, Neptune is much farther away from Earth and therefore offers a much smaller apparent disk to eclipse a star. An occultation by Neptune happens only about once a year.

Second, the axial tilt of Uranus helps the chances that its rings will occult a star, while Neptune's tilt does not. Planetary rings lie around their planet's equator. So, in the case of Uranus, for about half of its 84-year orbit, the rings present themselves face on to Earth, increasing the zone in the sky where a stellar occultation can take place. But Neptune stands closer to vertical in its orbit and thus its equatorial rings appear more nearly edge on, reducing the area of the sky they cover and hence the chance they will occult a star.

In 1981, Neptune nearly eclipsed a star, and a team of astronomers was watching intently. The star blinked once as Neptune approached but did not blink again after Neptune glided by. The astronomers announced a probable new moon of Neptune, but they weren't too confident. It wasn't a ring, they reckoned, because a ring would have occulted the star inbound and outbound. Yet for a moon, a very tiny dot, to partially occult a star was exceedingly improbable. Still, there are a lot of stars and such events must happen sometime. And something had indeed happened. But the suspected moon could not be seen directly.[5]

The search for a ring system continued. Neptune occulted a star in 1983 with many astronomers in attendance. No evidence of a ring or third moon was found.

Interested in the problem, an astronomer at the University of Paris, André Brahic, asked for telescope time on July 22, 1984, at the European Southern Observatory on Cerro La Silla, Chile, when Neptune was scheduled to nearly occult another star. His request was denied. But he drew attention to the event and prevailed upon some colleagues who had been assigned to the 40-inch (1-meter) telescope that night to take a break from their research and watch Neptune as it passed the star. There was a flicker before Neptune reached the star—but no flicker as Neptune moved away. The event had lasted about one second. The star's light had been dimmed 35 percent. It was as if there was a ring at Neptune but one that extended only part way around the planet—a partial ring, an arc.

Because of the attention Brahic called to the event, another team of astronomers was also at work that night using the passage of Neptune near the star to search for a ring. This team was only 60 miles (100 kilometers) south of Cerro La Silla, at the Cerro Tololo Inter-American Observatory. Their 36-inch (0.9-meter) telescope carried a sensitive photometer that continuously recorded the light levels on magnetic tape.

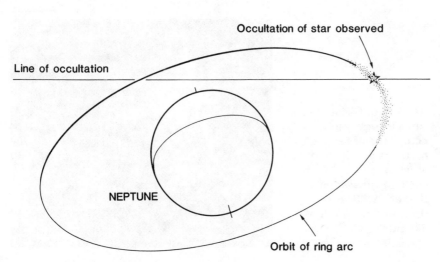

The arc ring system of Neptune was discovered in 1984 when Neptune passed near but not in front of a star. Yet the star was briefly occulted on one side of Neptune but not the other. Astronomers concluded that a thin ring was present but that it extended only part way around Neptune—a ring arc.

For quick reference, it also printed the average light level every 3.4 and 5.0 seconds. The occultation had lasted only about one second. The samples printed by the computer had so nearly averaged out the event that it went undetected.[6] But when the team leader, William B. Hubbard of the University of Arizona, heard about Brahic's positive occultation results at a conference that October, he reexamined the full record of the observation on his magnetic tape, which showed light-level readings every one-hundredth second. The occultation was real and agreed in every detail with the event seen at the European Southern Observatory. And because the same event had been seen by two observing teams at different sites, the occulting object could not be a moon. To partially block the same star simultaneously from slightly different angles would have required no less than two moons positioned for simultaneous grazing occultations with preposterous accuracy. The likelihood of such a coincidence was vanishingly small.

But a partial occultation seen from two sites could be explained by a short narrow stream of debris circling Neptune. Based on the length of the occultation, the ring arc was estimated to be about 15 miles (25 kilometers) across and 100 miles (160 kilometers) long. It was 41,500 miles (67,000 kilometers) from the center of Neptune. At 2.65 Neptune radii from the planet's center, this ring arc is near but beyond the planet's Roche Limit, the distance from a planet within which a large body cannot form due to tidal forces exerted on it by the planet. Because this ring arc lies beyond the Roche Limit for Neptune, the arc is probably the result of the breakup of a moon rather than primordial material that could not form into a moon.

Based on the number of searches for a ring that found nothing, Hub-

Artist Paul DiMare pictures a ring system composed of large dark particles and little dust or ice. *Voyager 2* found such a ring system at Uranus. Will the arc rings of Neptune look like this?
© Paul DiMare

bard, Brahic, and their collaborators estimated that the ring arc must envelop only about 10 percent of the planet.[7]

The discovery of an arc ring at Neptune provided continued stimulation for occultation observations at every opportunity. Approximately one-quarter of those experiments have produced readings that may be ring fragments, but according to researcher Philip D. Nicholson, only three results are absolutely convincing—and they yield three separate arc rings at different distances from Neptune. So there appear to be at least three arc rings, and there may be dozens. To date, none of the three best-established arc rings has been seen more than once. The first and outermost to be discovered lies 26,500 miles (42,700 kilometers) above the cloud tops of Neptune. The distances of the inner arcs are highly uncertain.

It was another hard-won victory to wrest information from a tiny blue-green dot almost 3 billion miles (4.8 billion kilometers) away.

But how had partial rings come to be, and how could they persist? The particles closer to Neptune must revolve faster, tending to spread the arc's inner boundary forward along its orbit. The particles farther from Neptune must revolve more slowly, tending to spread the arc's outer boundary backward along the orbit. Jack Lissauer calculated that rotational shearing would spread the particles of the arc into a full ring in less than three years. So what was preventing the arc or arcs from forming a full ring?

Several theoreticians have attempted to explain how an arc ring system could last millions of years, perhaps since the formation of the solar system.[8]

Peter Goldreich, Scott Tremaine, and Nicole Borderies offered a mechanism by which one shepherd moon could by resonances stabilize a series of arc rings if the moon's orbit was circular but slightly inclined to the arc ring plane. The mechanism is analogous to the moons Janus and Epimetheus in the Saturn system, which occupy almost precisely the same orbit. As Epimetheus, the smaller of the two, catches up with its companion, it is accelerated by the gravity of Janus and begins to ascend to a slightly higher orbit. As it rises, its orbital velocity declines and it begins to fall behind. In this way Epimetheus never passes and never quite catches up with Janus.

As Epimetheus falls behind, Janus gradually, over a period of years, begins to approach it from behind. The entire process repeats itself, this time with Janus falling behind and Epimetheus gradually catching up. In this way Janus and Epimetheus oscillate closer and farther from one another along the orbit they share.

This same process, say Goldreich, Tremaine, and Borderies, is probably at work in the Neptunian system, except that Epimetheus is replaced by a swarm of particles. Each particle in the arc rings behaves like a tiny moon and exhibits the same kind of oscillation as Epimetheus, catching up with and falling behind its coorbiting moon, which thus acts as shepherd. Astronomers have not yet seen that shepherd moon in the Neptunian system, but perhaps *Voyager 2* will.

Depending on its size, this coorbital shepherd would be able to explain not only a ring arc or arcs at about its own distance from Neptune but also, by resonance, a large number of arc rings at different distances.

According to their calculations, these arc rings would be stable for billions of years, so the partial rings could have been in place since about the time that Neptune formed.

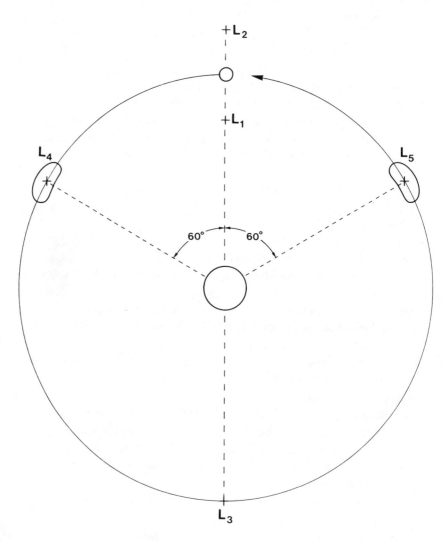

Lagrangian Points. For any two bodies revolving around one another, there are five precise positions where the gravity of the two bodies and centrifugal force will allow small objects to remain. Only the L4 and L5 points are truly stable, however, for small objects disturbed from those points will oscillate near those positions and not fall toward either body.

A different method of sustaining Neptune's peculiar ring arc system has been proposed by Jack Lissauer using a fascinating arrangement of shepherd moons. To explain how the arc particles stay together in a clump along their orbit, Lissauer took his cue from Jupiter. As Jupiter revolves around the Sun, two families of asteroids travel with it in its orbit, one 60 degrees ahead of Jupiter and another 60 degrees behind. These minor planet families of Jupiter are called the Trojan asteroids, and their orbital curiosity was first explained by the Italian-French mathematician Joseph Louis Lagrange in 1772, more than a century before asteroids began to be discovered at the Lagrangian Points.

At these positions the gravity of Jupiter and the gravity of the Sun balance in such a way that particles and moonlets located there remain equidistant from the Sun and Jupiter with stable orbits. Three other gravitationally balanced positions also exist for any body and its orbiting satellite. But of the five, only two provide stable orbits: those that lie along the satellite's orbit and precede and trail it by 60 degrees. The preceding Lagrangian Point is designated L4; the trailer L5.

At Jupiter, the Sun and planet combine to produce gravitationally stable positions for the Trojan asteroids. Lissauer used that scenario for Neptune with the substitution of the ring arc for the Trojan asteroids and

LAGRANGIAN POINTS

In 1772 mathematician Joseph Louis Lagrange discovered a peculiar consequence of the law of gravity when applied to two bodies orbiting one another, such as Jupiter and the Sun. According to his calculations, there would be two positions along the orbit of Jupiter 60 degrees ahead and 60 degrees behind the planet where the gravity of Jupiter and the Sun would cause small objects near those sites to oscillate without falling toward Jupiter or the Sun.

In 1904, 132 years later, asteroids began to be found at those positions, called the Lagrangian Points.

There are a total of five Lagrangian Points for any two bodies revolving around one another. The other three lie along a line connecting the two bodies, as shown in the diagram. These points, however, are not stable for small objects because if the objects are disturbed even slightly from those positions (by the Sun or another planet), they will wander away and not return.

No asteroids have been found at the Lagrangian Points in the Earth-Sun or Earth-Moon systems, but there have been reports of particles detected at the orbital positions 60 degrees ahead (L4) and 60 degrees behind (L5) each of the smaller bodies.

No asteroids have been found at the Lagrangian Points in the Saturn-Sun system either, but the gravities of Saturn and its modest-size moon Tethys work together to hold two moons at Lagrangian Points along Tethys' orbit. Calypso occupies the L4 position and Telesto follows at L5. Both Lagrangian moons are about 15 miles (25 kilometers) in diameter.

Saturn appears to delight in Lagrangian possibilities. The ring-master and its moon Dione combine gravitational forces to hold Helene (about 20 miles in diameter) at Dione's L4 point.

the substitution of Neptune and an undiscovered moon for the Sun and Jupiter. He suggested that the undiscovered moon was 60 degrees behind the arc ring (which would place the arc ring in the L4 position). To accomplish its gravitational duties, this Lagrangian shepherd would have to be about 125 miles (200 kilometers) in diameter—a candidate for *Voyager 2* discovery because it is too small to be seen from Earth.

The Lagrangian shepherd would keep the arc ring particles from spreading out along their orbit around Neptune but would not control the spread of the particles outward and inward into a wide band due to collisions that would accelerate some into higher orbits and decelerate others into lower orbits. Eventually these collisions would propel particles far enough from the stable Lagrangian Points so that the arc rings would be extremely tenuous. So at least one more shepherd moon was necessary to control the radial spread of the arc. Like the shepherds for the full rings at the other gas giants, this shepherd could orbit inside or outside the arc ring. There could be an inner shepherd, an outer shepherd, or both.

Lissauer was forecasting that this shepherd would lie inward of the ring where its greater orbital velocity would cause it periodically to catch up with and pass the arc ring particles and accelerate descending (closer) ones slightly so that they would climb back into the arc ring clump and not decay into Neptune. There is no way to know the size of this shepherd because the required mass would vary according to how far its orbit is from the partial ring. It is probably between 60 and 125 miles (100 and 200 kilometers) in diameter. A shepherd orbiting 600 miles (1,000 kilometers) below the orbit of the arc ring, said Lissauer, would need to be about 75 miles (120 kilometers) in diameter.

Which model, if either, is correct? Lissauer advises that the Goldreich-Tremaine-Borderies hypothesis deserves to be favored because it is more economical in the number of moons it requires.[9] Each ring arc in Lissauer's system would be stabilized by two or three shepherd moons, whereas in the Goldreich-Tremaine-Borderies explanation one moon would suffice to stabilize one or many arcs. If *Voyager 2* can photograph the expected shepherd moon or moons of Neptune in or near the arc ring system, it should be possible to settle this question.

Scientists are also anxious to see if *Voyager 2* will be able to photograph the ring arcs themselves. It won't be easy. As measured from their brief occultations of stars, the arcs are certainly very narrow, like the rings of Uranus, and probably very dark as well. With its ability to make closeup pictures and to provide its own radio occultation experiment, *Voyager 2* may be able to reveal the number and structure of Neptune's curious ring fragments and the size and nature of the particles that compose them.

VIEW FROM EARTH

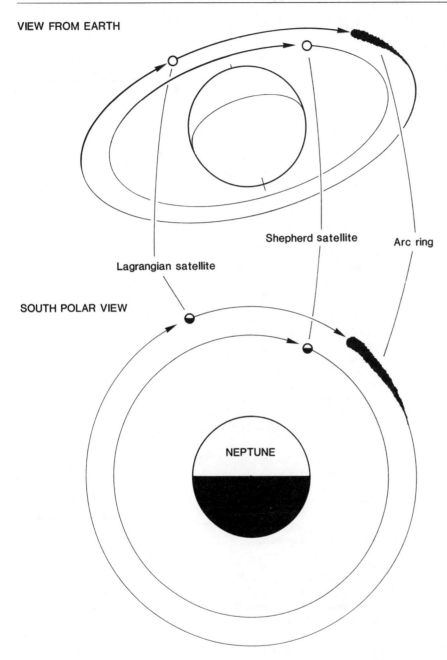

In the hypothesis by Jack J. Lissauer, each of Neptune's arc rings is maintained by the teamwork of two shepherd moons unseen from Earth. The arcs lie at the leading or trailing Lagrangian Points of one moon that keeps the fragments bunched up along their orbit. A second moon inside or outside the ring-arc orbit keeps the fragments from spreading inward or outward.

Shepherd satellite

Arc ring

Lagrangian satellite

SOUTH POLAR VIEW

NEPTUNE

Sixteen moons have been confirmed for Jupiter. Seventeen for Saturn. Fifteen for Uranus. Only two for Neptune.

Triton was discovered by the English brewer and amateur astronomer William Lassell soon after Neptune was found and was named ap-

The Moons of Neptune

SCORECARD FOR *VOYAGER 2* DISCOVERIES AT NEPTUNE

	Best Information *before* Voyager 2 *Encounter*	*Data from* Voyager 2 *Encounter* *(fill in)*
Neptune		
Diameter	30,800 miles (49,600 kilometers)	
Mass (Earth=1)	17.2	
Density (water=1)	1.76	
Rotation period	17.0 hours (based on clouds); about 18.0 hours suspected for interior	
Magnetic field	not detectable from Earth; suspected to be similar to Uranus' in strength	
Satellites	2; shepherd moon(s) for ring arcs expected to exist	
Ring arcs	3 reasonably certain; material fills about 20 percent of orbit; ring particles expected to be black	
Distance from center of planet	25,000 to 44,000 miles (41,000 to 71,000 kilometers)	
Distance above cloud tops	10,000 to 29,000 miles (16,000 to 46,000 kilometers)	
Atmospheric composition	hydrogen, helium, methane	
Atmospheric features	methane clouds barely discernible away from equator; slight banding expected but not seen from Earth	
Triton		
Diameter	2,200 miles (3,500 kilometers) (highly uncertain)	
Mass (Earth=1)	0.016	
Density	2 (highly uncertain)	
Rotation period	5.9 days retrograde	
Atmospheric composition	methane, nitrogen	
Atmospheric features	pressure and transparency unknown	
Surface features	none known; methane and water frost detected; possible nitrogen ice and/or liquid; dark reddish color expected due to methane polymerization	
Nereid		
Diameter	200 miles (300 kilometers) (highly uncertain)	
Mass (Earth=1)	negligible	
Density	2 (highly uncertain)	
Rotation period	365.2 days	
Atmosphere	none	
Surface features	none known	

propriately after the son of Poseidon (Neptune), god of the sea. He and his brothers, the tritons, had the upper bodies of men and the lower bodies of fish—the male version of mermaids.

A second satellite of Neptune was not discovered for more than a century. It was found in 1949 by Gerard P. Kuiper. Probably less than 200 miles (about 300 kilometers) in diameter, this small moon was named Nereid after the sea nymphs who served as attendants for Poseidon.

So, for the moment, Neptune is only known to have two moons. *Voyager* may change that. It tripled the number of moons known at Uranus. Neptune's greater distance makes the detection of moons smaller than 150 miles (250 kilometers) very difficult from Earth with ground-based telescopes. And it would be hard to explain Neptune's arc rings without invoking at least one shepherd moon of respectable size.

There could be surprises. What they lack in number, the moons of Neptune make up in weirdness.

Triton, the larger and closer of Neptune's two moons, travels around its planet in a retrograde orbit—circling in the direction opposite to Neptune's orbit around the Sun and rotation on its axis. Triton is the only large satellite in the solar system with this peculiarity.

Nereid is peculiar in its own way. It possesses the most eccentric orbit of any moon in the solar system, passing as close to Neptune as 870,000 miles (1.4 million kilometers), then swinging out to 6.0 million miles (9.7 million kilometers).

Unlike most other satellites, Triton and Nereid do not revolve around their planet's equator but instead occupy highly inclined orbits.

These anomalies combine to suggest to most planetary astronomers that something catastrophic has happened to the satellite system of Neptune. Perhaps the disruption and damage have been so great that the number of moons waiting to be discovered at Neptune is low—maybe only those required to hold the ring arc system in place.

Triton is big—but exactly how big is uncertain. It may be about 2,200 miles (3,500 kilometers) in diameter. That would make it considerably larger than Pluto and about the same size as our Moon. Of the satellites in the solar system, Triton may rank between fifth and seventh in size, behind Jupiter's Ganymede, Saturn's Titan, Jupiter's Callisto and Io, and perhaps behind Earth's Moon and Jupiter's Europa. Even if Triton has a larger diameter than our Moon, it probably has less mass because our Moon is denser.

Of all the 54 known satellites in the solar system, only two—Triton and Titan—possess permanent atmospheres. Infrared studies of Triton by Dale P. Cruikshank and Peter Silvaggio in 1975 identified methane ice on Triton's surface. That led to the realization that Triton must possess an atmosphere. It was not possible to distinguish methane gas from methane ice in the spectrum, yet at the temperature of Triton (about −360°F; 55°K), there had to be enough vaporization of methane ice to provide Triton with at least a very thin atmosphere.

Cruikshank and his colleagues also discovered on Triton the presence of molecular nitrogen (N_2). Much of it also would lie condensed on the surface, but enough would be vaporizing to give Triton a substantial atmosphere, with approximately one-tenth the surface pressure of the air on Earth.

Thus the atmosphere of Triton may be very similar to what the *Voyagers* found at Titan: a gaseous blanket dominated by nitrogen with a trace of methane.[10] Ultraviolet wavelengths of sunlight and bombardment by charged particles in Neptune's magnetosphere could cause methane in Triton's atmosphere to polymerize into smog, as it has at Titan. But Triton's atmosphere is so much less dense than Titan's that the smog could perhaps drop out of the sky, leaving a rich reddish sludge of organic chemicals on the surface on Triton and far less smog in the atmosphere.

The colder temperatures at Triton could indeed produce a bizarre landscape. In identifying methane ice on its surface, Cruikshank and Silvaggio noticed that as Triton rotates in a six-day period, the signal varies so as to indicate that the methane ice does not cover everything. Instead the methane ice seems to be distributed in continent-size patches, between which water ice, rock, soil—and nitrogen—are exposed. The temperature on Triton might allow the nitrogen to be frozen or liquid. Cruikshank

TIMELINE: *VOYAGER 2* AT NEPTUNE

Neptune encounter

Closest approach	9:00 P.M. PDT, August 24, 1989
	(4:00 A.M. GMT, August 25, 1989)
Distance	18,133 miles (29,183 kilometers) from center of planet
	2,700 miles (4,400 kilometers) above cloud tops
	420 miles (680 kilometers) above suspected atmospheric drag dangerous to *Voyager 2*

Triton encounter

Closest encounter	2:14 A.M. PDT, August 25, 1989
	(9:14 A.M. GMT, August 25, 1989)
Distance	25,000 miles (40,000 kilometers) from center of Triton
	23,900 miles (38,250 kilometers) above surface of Triton

Nereid encounter

Closest approach	5:12 P.M. PDT, August 24, 1989
	(12:12 A.M. GMT, August 25, 1989)
Distance	2.9 million miles (4.7 million kilometers)

envisions shallow lakes, a few feet deep, of liquid nitrogen. Some astronomers are skeptical because the temperature range for nitrogen as a liquid is very narrow. But it would be fascinating if Triton, son of the sea god Neptune, had such peculiar oceans on its surface.

Will *Voyager 2* glimpse this landscape? Is the atmosphere of Triton transparent? Over the last few years the light variation from Triton as it rotates has virtually vanished. Scientists worry that as Triton approaches its summer season, its atmosphere is becoming hazy. Still, *Voyager*'s spectrometers and the analysis of its radio signals passing through the atmosphere of Triton should provide conclusive information about the gases that blanket this moon.

If Triton's atmosphere is transparent, *Voyager 2* may indeed see a moon to rival Jupiter's Io, Saturn's Titan, and Uranus' Miranda in spectacle and mystery. Triton must have led a remarkable life. It is the only large moon in the solar system to revolve in a retrograde direction. Triton must have undergone a major gravitational ordeal after it was formed. Perhaps *Voyager 2* will see Triton's history written on its surface features.

Late on August 24, 1989, *Voyager 2* will pass closer to Neptune than it did to Jupiter, Saturn, Uranus, or any of their moons or rings—only about 2,700 miles (4,400 kilometers) above its cloud tops as it skims over Neptune's north pole[11] and heads on for Triton and then the stars. *Voyager 2* will have completed its Grand Tour of all four giant planets in our solar system.

CHAPTER 11

Voyagers to the Stars

"This is a present from a small distant world . . . We are attempting to survive our time so we may live into yours. We hope someday, having solved the problems we face, to join a community of galactic civilizations. This record represents our hope and our determination, and our good will in a vast and awesome universe."

From President Jimmy Carter's message to extraterrestrial civilizations on the *Voyager* spacecraft record (1977)

Voyager 2's Grand Tour of the outer planets must end with Neptune in 1989. To obtain a gravity assist from Neptune so that *Voyager* could travel on to Pluto would require the spacecraft to fly within 3,100 miles (5,000 kilometers) of the center of Neptune. Since the radius of Neptune is 15,400 miles (24,800 kilometers), this extension of the mission is not practical.

But *Voyager 2*'s mission will not end with the Neptune encounter. Both *Voyager 1* and *Voyager 2* received enough energy from gravity assists at Jupiter to escape from the solar system. Their subsequent planetary encounters have been along trajectories that will permit them to escape the gravitational bonds of the Sun and send them out of our solar system on an endless glide through interstellar space.

The Journey Beyond

But crossing the orbit of Neptune or Pluto does not place the spacecraft beyond the solar system. Beyond the outermost planets lies the Oort Cloud, a reservoir of trillions of comets, extending halfway to the nearest star. Its outer boundary is perhaps 7,000 times the distance of Pluto from

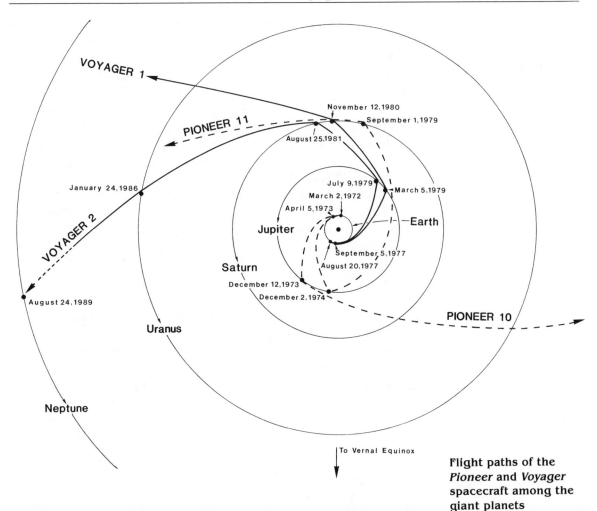

Flight paths of the *Pioneer* and *Voyager* spacecraft among the giant planets

the Sun, yet the comets of the Oort Cloud are held to the solar system by the Sun's gravity. For *Voyager 1* and *Voyager 2*, leaving the solar system is a threshold that lies far in the future.

Yet there is a boundary that both *Voyager* craft may survive to mark. The *Voyagers* are sailing outward with the solar wind at their backs—a breeze of electrons and protons from the Sun racing past them. Near the Earth, the solar wind regularly blows at a million miles per hour (500 kilometers per second). But somewhere, well beyond Neptune, the speed of the solar wind falters as the outbound particles are confined by the pressure of the interstellar gas and magnetic field through which the solar system swims.

This bubble of confinement is the heliopause—effectively the end of the Sun's far outer "atmosphere"—the point where the environment is less that of the Sun than of the combined outpourings of all the other stars in the Milky Way Galaxy. It is an important threshold as the *Voyagers* wend their way toward the stars.[1]

The location of the heliopause—its distance from the Sun—is unknown. No spacecraft escaping from our solar system has yet reached this boundary. Only four spacecraft are outbound with escape velocities. At the beginning of 1988, *Pioneer 10* was about 42 astronomical units from the Sun, *Pioneer 11* was 24, *Voyager 1* was 31, and *Voyager 2* was 24. Thus, both *Voyagers* are already farther from the Sun than *Pioneer 11*, which reached Saturn via gravitational assist from Jupiter by traveling more across the solar system than directly outward. *Voyager 1* will overtake *Pioneer 10* in distance in 1998. *Voyager 2* will overhaul *Pioneer 10* in 2016.[2]

Pioneer 10 is leaving our solar system in a direction nearly opposite to the Sun's motion among the stars. The distance to the heliopause upstream (in the direction the Sun is moving) is expected to be shorter than the distance downstream because the speed of the Sun in the upstream direction causes the pressure of the interstellar medium to be greater, pushing the heliopause inward. Therefore, *Pioneer 10*, headed downstream, may have farther to travel to reach the heliopause than the other craft.

The distance to the heliopause must also be somewhat variable because of the changeable output of the Sun. Faster solar particles push the heliopause outward, while periods of slower-moving particles allow the

---

The top diagram on the facing page shows the flight paths of the *Voyager* and *Pioneer* spacecraft as they leave the realm of the planets as viewed from north of the solar system. The bottom diagram shows the *Voyager* and *Pioneer* trajectories from the side, 20 degrees above the plane of the solar system, viewed from the vernal equinox. The end of each spacecraft's trajectory marks its position in the year 2000 A.D. Here are the directions (based on the ecliptic) that the spacecraft are headed and their asymptotic speeds.

Spacecraft	Asymptotic Speed (astronomical units per year)	Celestial Latitude	Celestial Longitude
Pioneer 10	2.384	2.909°	83.368°
Pioneer 11	2.214	12.596°	291.268°
Voyager 1	3.501	35.549°	260.778°
Voyager 2	3.386	−47.455°	310.885°

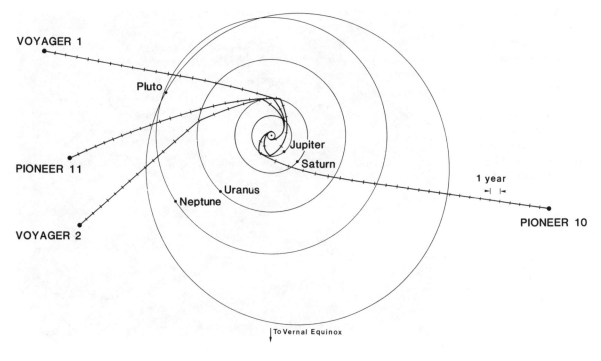

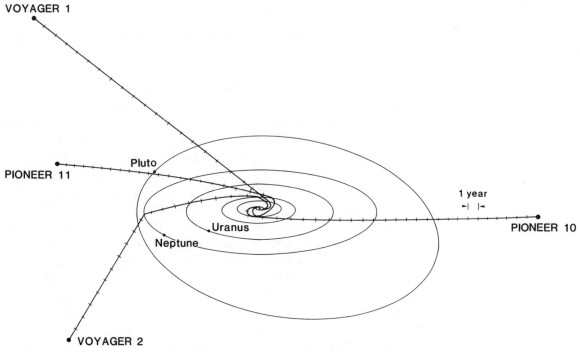

NASA's *Pioneer 10* and *11* and *Voyager 1* and *2* spacecraft are headed out of the solar system. Artist Don Davis portrays *Pioneer 10,* at present the most distant craft, moving outward from the Sun (the brightest star visible) and away from the heart of our Milky Way galaxy.
NASA Ames Research Center

heliopause to contract. Solar scientists think that the heliopause lies at a distance of 100 to 200 astronomical units—100 to 200 times the Earth's distance from the Sun. Neptune orbits at a distance of 30 astronomical units.

Voyager 1 and *Voyager 2*, increasing their distance from the Sun by about 3.5 astronomical units per year, could arrive at the heliopause as early as about 2010. Both craft should continue to have enough electrical power from their nuclear generators for radio transmission and enough fuel for their thrusters to keep their antennae turned toward Earth until about 2010 or 2015. If the heliopause lies at about 100 astronomical units from the Sun, the *Voyagers* may tell us where. Beyond that point, the *Voyagers* will continue outward, but they will be mute and passive.

Yet even at the heliopause, the *Voyagers'* journey of departure from the solar system will scarcely have begun. Beyond them will lie the vast cloud of comets, no less members of the Sun's family than the planets. The Oort Cloud may begin as close to the Sun as 50 astronomical units and extend as far as 135,000, halfway to the nearest star. Even though the *Voyagers* are separating themselves from the Sun at about 10 miles per second (16 kilometers per second), it will be 40,000 years until they are beyond the Oort Cloud and have truly crossed the boundary into interstellar space. Forty thousand years is about the period of time that separates us from Neanderthal Man.

On its Grand Tour of the giant planets, *Voyager 2* has been traveling near the plane of the solar system. But at Neptune, *Voyager 2* will pass low over the planet's north pole, and its course will be bent southward to encounter Triton. So *Voyager 2* will be headed out of our solar system on a southerly route. It is pointed toward a rather drab region of the sky in the constellation Pavo, the Peacock. To see that part of the heavens, we must be in the southernmost United States or farther south.

Voyager 2 and the Stars

Yet, ironically, *Voyager 2*'s first reasonably close stellar encounter will not be with any of the stars in the far southern sky but with a star now located in the northern constellation Andromeda, the Princess. That star, known only by its catalog designation of Ross 248, is a small cool red star with only about one-fifth the mass of our Sun. Find the Great Square of Pegasus and look north one length of the Square—and you will not see Ross 248. With a magnitude of +12.3, it is about 200 times too faint for human eye visibility.

Even though *Voyager 2* will be traveling south and Ross 248 is located in the north, the two are moving toward one another. While *Voyager 2* is consuming 40,000 years in its transit through the Oort Cloud, all the stars in the sky are moving in different directions. The constellations are gradually changing shape beyond recognition. Some of the stars are coming toward us faster than our spacecraft are going to meet them—almost as if they are coming to fetch the *Voyagers.*

Ross 248 is currently 10.3 light-years away, but while *Voyager 2* rushes outward at 33,000 miles per hour (14.8 kilometers per second), Ross 248 is approaching our system at more than five times that speed. No sooner will *Voyager 2* emerge from the Oort Cloud than, 40,176 years from now, it will encounter Ross 248, passing at a distance of 1.7 light-years, closer to *Voyager* than to any subsequent star known. Ross 248 will pass the outskirts of our solar system 3.25 light-years from our Sun, 25 percent closer than our nearest stellar neighbors, the three stars of Alpha Centauri, are to us now. Yet even when Ross 248 reaches that close range, it will be four times too faint for people on Earth to see without a telescope.

Still, its passage may eventually be seen and even felt indirectly as its gravity warps the orbits of millions of comets and redirects some of them inward toward the Sun where they will provide brilliant displays in the skies of Earth and perhaps even impacts on our planet.

Voyager 1 and the Stars

The same kind of stellar encounter awaits *Voyager 1*, even though it is traveling toward a very different part of the sky. It is pointed in the direction of Rasalhague, the brightest star in the constellation Ophiuchus, the Serpent Bearer. But the star headed for a rendezvous with *Voyager 1* is AC+79 3888.[3] This star, with no name other than its catalog listing, is currently to be found in the faint constellation Camelopardalis, the Giraffe. This region of the sky is visible all night long to people living north of the Tropic of Cancer. AC+79 3888 is just a short distance from Polaris, the North Star, and halfway between the bowl of the Big Dipper and the *W* of Cassiopeia. AC+79 3888 is slightly larger and brighter than Ross 248, but it too is a small cool red star with only one-quarter the mass of our Sun. At its present distance of 16.6 light-years, AC+79 3888 is an 11th-magnitude star, nearly a hundred times too faint for the unaided eye to see.

While *Voyager 1* is moving outward at 37,000 miles per hour (16.6 kilometers per second), AC+79 3888 will be traveling toward our solar system at seven times that speed. In 40,272 years, at the same time that *Voyager 2* will be scurrying by Ross 248 more than a quarter of the way around the sky, *Voyager 1* will be only a little more than 1.6 light-years from AC+79 3888, and AC+79 3888 will be just 3 light-years from the Sun.[4] Even so, AC+79 3888 will still be two times too faint for people on Earth to see without a telescope.

A Message from Earth

Quite by accident, both of these first star encounters by the *Voyagers* are with single stars like our Sun—a minority in space, where most stars have one or more gravitationally bound companions. They may well have planetary systems, since we think the process that starts the formation of stars is identical to the process of planet formation.

Yet even with a family of planets, it is very unlikely that Ross 248 and AC+79 3888 provide the right environment for life to exist. Both stars are much smaller than our Sun. They emit so little heat that a planet would have to be at precisely the correct distance with an almost perfectly circular orbit to stay in a habitable zone. Worse still, that planet would be so close to its star that it would be tidally coupled to it, like most moons are to their nearby planets, so that one side of the planet would fry in constant sunlight while the other side would freeze in constant night. Most scientists do not expect life to exist in the solar system of a low-mass red dwarf star.

Even if both these stars illuminate planets populated by intelligent spacefaring beings, it would be extremely unlikely that they would detect a tiny silent spacecraft passing beyond the fringe of their comet clouds.

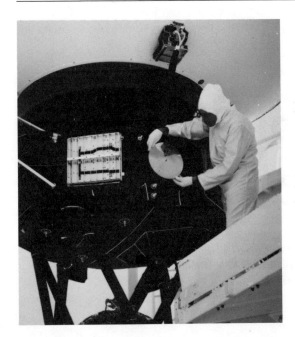

Voyager 1 and *Voyager 2*, bound out of the solar system, carry sounds and pictures of Earth on a phonograph record to show a civilization that may find the spacecraft what life on our planet is like. Here, a technician is mounting the interstellar message on *Voyager 2.*
NASA

And the stars are so widely separated that there is a vanishingly small chance that either of the *Voyagers* will hit or come very close to a star in the next billion years.

Still, just on the outside chance that some civilization deep in space may retrieve a *Voyager*, each craft is equipped with a special record that gives its finders pictures and sounds from the planet Earth.

The message was designed for NASA by Carl Sagan, Frank Drake, Ann Druyan, Timothy Ferris, Jon Lomberg, and Linda Salzman Sagan. Attached to the side of each *Voyager* is a gold-coated two-sided copper phonograph record, complete with enclosed stylus and cartridge and with instructions etched on its aluminum cover. The record should last a billion years. On it are greetings in 55 different human languages and one whale language; the sounds of Earth—from thunder to frogs to a newborn baby; 90 minutes of music from around the world; and, encoded as vibrations, 118 pictures of our planet and ourselves.[5]

For the beings that find a *Voyager* along its endless journey, the spacecraft will have found a new and eloquent voice—no longer telling its home planet about other worlds but now telling other beings of its origin and the people who sent it outward.

CHAPTER **12**

The Smallest Planet

"One cannot help but wonder whether this remarkable pair (Neptune and Pluto) holds any further surprises."
Astronomer James G. Williams
(1971)

The Incredible Shrinking Planet

Pluto was hard to find. Of the few who sought to locate a trans-Neptunian planet, most justified their efforts by what seemed to them to be slight but real irregularities in the motion of Uranus even after the perturbations of Neptune were accounted for. Pluto was finally discovered in 1930, culminating a series of searches that began in 1905. But even after its discovery, Pluto, like its mythological namesake, was a shadowy figure, if not downright antisocial. The planet revealed almost nothing to astronomers.

No moon was seen, so it was not possible to calculate Pluto's mass using the law of gravity.[1] Pluto's disk as seen from Earth was too small to measure, so it was not possible to determine the diameter of the planet with any accuracy at all. The planet revealed no surface features. The dimness of the planet and the inability of existing telescopes to see its disk meant that, unless Pluto was absurdly dense, it was too small to be the planet that Lowell and Pickering had predicted. Its mass could not have measurably disturbed the motion of Uranus.

When it was discovered, Pluto was thought to be about the size of Earth. As telescopes grew and techniques improved, the planet stayed elusive, always small enough to avoid precise measurement. It was thought to be three-quarters our size, then half. In 1950, Gerard P. Kuiper tried to

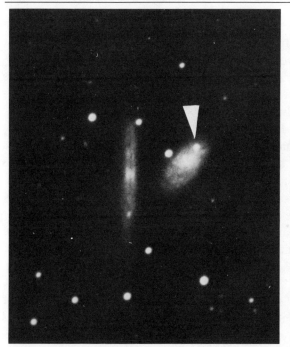 

Pluto wanders in front of two galaxies. In the picture on the left (October 28, 1970), Pluto is the starlike object at the top of the galaxy on the right. One day later (picture on right), Pluto has moved to a position where it appears between the two galaxies. James Gunn took the photograph on the left with the 48-inch Schmidt Telescope at the Palomar Observatory. The photo on the right was taken by Karen and Richard Hackney and Alex G. Smith with the 30-inch telescope at the Rosemary Hill Observatory, University of Florida.

(left) Palomar Observatory photograph by James Gunn; (right) Courtesy of Karen Hackney, Richard Hackney, and Alex G. Smith

measure Pluto's disk with the newly operational 200-inch telescope on Palomar Mountain and concluded that Pluto was less than 3,500 miles (5,900 kilometers) in diameter.[2]

In 1955, Merle F. Walker and Robert Hardie at the Lowell Observatory identified fluctuations in the light from Pluto recurring over a period of 6.3867 days. They interpreted this (correctly) as the rotation period of Pluto.[3]

Then, in 1976, astronomers[4] used spectroscopy to identify frozen methane on Pluto's surface, and estimates of the planet's size plunged again. Until that time it had been thought that Pluto was a dark-colored object with low reflectivity like our Moon or Mars or the asteroids. But with methane ice on its surface, Pluto was now thought to be a highly reflective body. If Pluto was reflecting 40 to 50 percent of the light it received yet still was very dim, it had to be even smaller than previously estimated, or more light would have reached the Earth. Pluto was smaller than Mercury. It had fallen to last place in size among the planets in our solar system. Pluto was probably smaller than our Moon.

The existence of methane ice on Pluto led astronomers to the realization that Pluto must have at least some atmosphere. Methane freezes only at extremely low temperatures. At Pluto, where temperatures are 45 to 55° above absolute zero (−379 to −361°F; −228 to −218°C), methane is frozen but is steadily vaporizing to supply Pluto with an atmosphere. A temperature change of just a few degrees changes the rate of vaporization and the amount of atmosphere greatly. The behavior of methane makes it certain that Pluto has a tenuous atmosphere of methane, but it is very hard spectroscopically to distinguish methane gas from methane ice.

On June 9, 1988, Pluto edged in front of a star, the first confirmed stellar occultation for tiny Pluto since it was discovered. The star's light faded and returned gradually rather than blinking out and then back on suddenly, indicating that Pluto is surrounded by a partially transparent medium—an atmosphere. Methane must be present, but the other and even the principal components of the atmosphere have not yet been determined. Whatever its composition, Pluto's atmosphere can't be extensive. The gravity of Pluto can't hold much.

Still, it is surprising that a planet of such small size—with so little gravity—can retain any atmosphere at all. The atmosphere is certainly not tightly bound to Pluto. The methane gas must be drifting away very gradually as solar energy, weak as it is at Pluto, accelerates some methane molecules to Pluto's escape velocity—a mere 2,700 miles per hour (1.2 kilometers per second). If methane molecules at the top of Pluto's atmosphere are traveling outward and do not collide with other molecules, they are lost. To keep Pluto supplied with its present methane atmosphere since the beginning of the solar system, the planet would have needed a global coating of methane ice at least 2 miles (3 kilometers) thick.[5] As skimpy as Pluto's atmosphere is, astronomers think it is many times thicker now than usual, as Pluto makes its closest approach to the Sun and the warming temperatures vaporize more methane ice. For all but a few years of the planet's 248-year circuit of the Sun, Pluto's atmosphere may lie frozen on the ground.

Measurements following perihelion may indicate whether or not the atmosphere of Pluto is transient. Certainly Pluto's maintenance of an atmosphere is precarious. If it were a little colder, Pluto's atmosphere would freeze to the planet's surface. A little warmer and virtually all of Pluto's wisp of an atmosphere would escape, unless continuously replenished by the evaporation of surface ices.

For nearly half a century after its discovery, Pluto had shrunk from view and defied efforts of familiarity. But gradually, hardworking astronomers (and Pluto's eccentric orbit) coaxed the planet into a more accommodating mood. Suddenly, everything Pluto did seemed almost to beg for attention. Pluto turned out to be something of an exhibitionist.

The Discovery of Pluto's Moon

The crucial breakthrough in Pluto research came on June 22, 1978, when James W. Christy was examining photographs of Pluto he had requested. They had been taken as part of an ongoing effort by the U.S. Naval Observatory to establish a more accurate orbit for the planet.[6] Pluto had been photographed by Anthony V. Hewitt on three different dates using the 61-inch (1.55-meter) Astrometric Reflector at the Naval Observatory's Flagstaff (Arizona) Station. There were two plates for each date and three images of Pluto on each plate. They did not look very promising. The pictures had been marked "poor" because the image of Pluto appeared to be defective. It was elongated. It would be hard to get measurements of Pluto's precise position from a dot that had been elongated asymmetrically as if the image had been smeared when the telescope failed to track it properly as the Earth turned during the 1.5-minute exposure.

But then Christy noticed that the stars around Pluto were not elongated. He looked again at the elongations of the Pluto image. The bulge had changed position. On the April 13 and 20 plates the elongation was to the south. On May 12, the elongation was to the north. Christy toyed with the idea that Pluto had a moon, but he remained skeptical. He discussed his findings with his colleagues, looking for other explanations, but none seemed to fit.

The next day, Christy examined the 50 earlier U.S. Naval Observatory plates of Pluto taken in 1965, 1970, and 1971. Two plates from 1965 showed the bulge. But most interesting were five plates taken during one week in 1970. They showed the elongation progressing clockwise around Pluto in a 6-day period, about the same as Pluto's rotational cycle.[7] Christy sought out his colleague Robert S. Harrington. "Bob," said Christy, "Pluto's got a satellite."[8]

Christy measured the elongation angles carefully so that Harrington could attempt a calculation of the orbit of the satellite. He began with the assumption that the moon's period of revolution was exactly the same

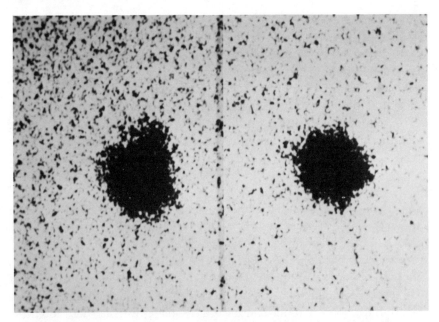

In 1978, James W. Christy noticed that there was a bulge in a highly magnified image of Pluto (left). He noticed that only Pluto, not the background stars (such as one shown to the right), exhibited this distortion. He found elongated images of Pluto on earlier U.S. Naval Observatory photographs. His conclusion: Pluto has a moon. He proposed the name Charon.

Official U.S. Naval Observatory photograph

as Pluto's period of rotation. Using a pocket calculator at home, Harrington computed the satellite's probable orbit and compared it to the elongations that Christy had measured on the photographs. The moon's predicted positions fit the image bulges almost perfectly. "That was great fun," said Harrington.

Pluto had a moon. It proved to be the largest satellite to be discovered since Neptune's moon Triton was found in 1846. No sooner had Harrington computed the moon's orbit than he used its orbit to derive the mass of the Pluto-satellite system. He also noted that Pluto and its moon were gravitationally bound together in a unique way. Moons typically revolve around their planets in the same period of time as they rotate once, so that they always keep the same face pointed at their planet. Unlike any other planet, Pluto also rotates once for every revolution of its moon, so Pluto keeps the same face pointed back at its moon.

A MOMENT OF PERCEPTION
by Dr. James W. Christy

From an apparently insignificant detail on a small number of astronomical images came a bonanza of scientific revelations. Pluto, a mystery for 48 years, became a well-understood planetary system within a span of a few days. How did this happen? Why was Charon not seen earlier by the thousands of astronomers who studied Pluto?

The elongation on the images of Pluto that was Charon was minute and hidden from perception by many effects: elongation due to the orbital motion of Pluto during exposure, elongation from imprecise tracking of the telescope during exposure, expansion of the image of Pluto due to poor "seeing" through the atmosphere, distortion of the image from imperfect telescope optics, and distortions from uneven distribution of photographic grains.

The elongation on the images of Pluto was also hidden by the bias in the minds of most astronomers, including myself, that moons did not exist that close to planets and, because many attempts to find a moon of Pluto had failed, that Pluto did not have a moon. Many of us had even been taught that. Thus, even though I had seen several exposures of Pluto with significant elongation, my mind was not yet open to discovery.

From the day that Pluto was discovered in 1930, astronomers had hoped to find a moon of the planet and thus calculate its mass. They made stronger exposures and Pluto's image overwhelmed that of Charon. They made underexposures in which the image of Pluto's companion was undetectable. Also, Pluto was at opposition (closest to Earth) in midwinter through most of the period. During winter, most observatories suffer from poorer "seeing" conditions and optical distortions, both due to thermal gradients in and around the telescopes. Pluto's orbital motion has caused opposition to slowly migrate until, at present, it occurs in the spring.

The discovery exposures of Charon were made in Flagstaff far from opposition in late spring near the end of the observing season for Pluto. Better

James W. Christy, discoverer of Charon, at the U.S. Naval Observatory's Starscan equipment, with which he determined that the bulge in Pluto was a moon.
Courtesy of James W. Christy

seeing conditions occur then in Flagstaff. The exposures were made with a telescope of high optical quality and low thermal coefficient optics. The exposure times just happened to be exactly right for detection of the elongation of the image, which was Charon.

When I first saw these exposures on June 22, 1978, I was looking with the mind and eyes of an astronomer who had examined roughly 50,000 images in recent years. Many of these images had been of double stars exposed in the course of the U.S. Naval Observatory's extensive double-star program. I had seen dual images blended together in all possible circumstances by all combinations of image distortions. My mind was now attuned to two celestial bodies disguised as one. Now I could think: Pluto has a moon.

In our plate collection, I found elongations of Pluto's images on three nights in the spring of 1978, on five nights in one week during 1970, and on two nights in 1965. The 1965 plate envelope had been clearly marked by the observer, "Pluto image elongated." I myself had seen that plate many times, but I had rejected the implication. Each of these nine nights revealed an elongation in a specific direction.

From ten apparently insignificant image distortions and two days of study came a cascade of revelations:

1. In the 1978 exposures, one night showed an elongation to the south; another, about a month later, to the north. Hypothesis: Pluto had a moon orbiting north-south. Strange—almost all other moons in the solar system move in an east-west plane. But no proof of a new moon yet.

2. In the 1970 exposures, five elongations in one week and the elongations progressed completely around Pluto. Conclusion: Pluto has a moon with a probable orbital period of six days.

3. But astronomers had previously determined a light variation of Pluto (presumably its rotation period) of precisely 6.3867 days. Hypothesis: The orbital period of the moon is exactly 6.3867 days.

4. Eureka! The two 1965 elongations, the five 1970 elongations, and the three 1978 elongations are at angles in exact agreement with the 6.3867 day period, according to calculations by Bob Harrington. How can we get so much from such trivial elongations? But the avalanche of conclusions continues.

5. The orbit is north-south with Charon only six Pluto diameters' distance from the planet. Harrington concludes that both bodies are tidally locked. This is also evi-

dent from the coincidence of the light curve and the orbital period. Thus, Pluto's pole of rotation is on its side, in the plane of the solar system. More yet.

6. Harrington calculates the mass of Pluto-Charon from the orbital period and the distance. He gets less than one percent of Earth's mass. Therefore, Pluto could not have have caused perturbations in the orbits of Neptune and Uranus.

7. Because a cold planet's density cannot be too low, the small mass leads to a very small diameter for Pluto, approximately 3,000 kilometers, we thought at the time. Charon's brightness is about 20 percent of Pluto's. Conclusion: Charon's diameter is more than a third of Pluto's, making it the largest moon in the solar system relative to its planet and the largest moon discovered since 1846.

8. During the time between the 1965 and 1978 elongations, Pluto's orbital motion around the Sun should change our view of Charon's orbital plane. The 1970 and 1965 elongations were extended farther east-west than those in 1978. The resultant orbital solution by Bob Harrington reveals that Charon's orbit was approximately face on at the time of Pluto's discovery in 1930 and has been closing to the line of sight since then. Conclusion: Eclipsing of Pluto-Charon will commence soon, even though this can occur only once every 124 years!

Thus within a week of Charon's discovery there were roughly a dozen major conclusions made concerning the true nature of Pluto. All of these conclusions have been verified over the last ten years.

What had been a prolonged mystery was quickly and totally unveiled by ten minuscule image distortions. By amazing coincidence, these ten elongations were ideally placed in time for scientific logic to yield correct conclusions.

As the discoverer of Pluto's satellite, it fell to Christy to suggest a name. He wanted to honor his wife Charlene—"Char" to her family. So he proposed the name Charon, the boatman in Greek mythology who ferries the souls of the dead across the River Styx to Pluto's realm in Hades.[9] The name was ideal: Charon as a sentinel for Pluto. It fit well with the lore of Pluto and the tradition of naming planets and moons after Roman and Greek mythological figures.

But the name Charon accidentally led to a problem with pronunciation. Christy had anticipated that Charon would be pronounced like "Char" in Charlene. But the standard Greek and Latin pronunciation of Charon is the same as the name Karen.[10]

Significance of a Moon for Pluto

The discovery that Pluto had a moon was not just fodder for trivia games and an excuse to revise textbooks. The existence of a satellite for Pluto vastly increased what could be known about Pluto from Earth-based observations.

To begin with, a satellite held in orbit by Pluto's gravity meant that scientists for the first time could calculate with some confidence the mass of the Pluto-Charon system.[11] Charon lay at a distance of 12,400 miles (19,980 kilometers) from Pluto and circled with a period of 6.39 days. From these figures, the mass of Pluto could be reasonably estimated. It would take about 450 Plutos to equal the mass of the Earth. Even our Moon is five and a half times more massive than Pluto.

Compared to Pluto, Charon was surprisingly large. Previously, the Earth held the honors for having the largest satellite in proportion to its size. The Moon was more than one-quarter the diameter of Earth and one-eighty-first its mass. No other planet-satellite pair came close to such equality in size. Earthlings talked proudly of their unique twin-planet system—until Charon's discovery. With an estimated diameter of 777 miles (1,250 kilometers) compared to Pluto's 1,457 miles (2,345 kilometers), Charon is more than half the diameter of Pluto and probably has about 15 percent its mass.

Lowell and Pickering had embarked upon their searches for a trans-Neptunian planet because they thought they had detected its perturbations in the motion of Uranus. They had used those irregularities (Pickering used what he thought were Neptune perturbations also) to predict the existence of a ninth planet. Driven by this work, Pluto was found, but its small size worried astronomers that its gravity was too weak to cause detectable perturbations in the motion of Uranus or Neptune. Yet almost no one had anticipated that Pluto had so little mass that the Earth from 3 billion miles (5 billion kilometers) away would exert a greater gravitational force on Neptune than did Pluto, even if Pluto and Neptune could be where their orbits lie closest together. Whether or not the oddities in the motion of Uranus and Neptune were real rather than errors of measurement, it was now absolutely certain that Pluto could not possibly be responsible. Its gravity was far too weak.

But Charon did far more than reveal the mass of Pluto. Before Charon was found, no one knew how Pluto's axis was tipped to its plane of revolution.[12] Did Pluto "stand" nearly vertically on its axis as it revolved around the Sun like seven of the planets, or did it lie on its side like Uranus? Charon, the bulge in the fuzzy image of Pluto, appeared alternately to the north and then to the south. If Charon was orbiting close to the plane of Pluto's equator as almost all moons in the solar system do, that meant that Pluto must have its equator tipped almost north/south as seen from Earth. For Charon to revolve precisely once around Pluto in the same period as Pluto spins once on its axis indicated that Charon must be positioned over Pluto's equator. So Pluto's rotational axis had to lie near the plane of its orbit. Pluto, like Uranus, lies on its side as it orbits the Sun.

The orientation of Charon's orbit also led to a likely explanation of why Pluto was not brightening as fast as expected as it approached perihelion. The closer it is to the Sun, the more light Pluto receives and the larger its angular disk is to reflect sunlight. Pluto should have brightened considerably between 1950 and 1987. It did brighten, but about 32 percent less than expected.

When Charon was discovered, its motion around Pluto showed that Pluto was tipped on its side as it revolved around the Sun. The oblique perspective on Charon's orbit at the time of its discovery in 1978 meant that when the ninth planet was discovered in 1930, Tombaugh must have been viewing the region around Pluto's south pole.[13]

In the late 1980s, with the gradual change of seasons, Pluto's equator is facing the Sun and Earth. The failure of Pluto to brighten as much as expected is thought to be due primarily to this change in perspective. If the poles of Pluto are covered with ice and highly reflective, while the equator is much darker (perhaps in part because it is exposed to almost vertical sunlight as Pluto comes closest to the Sun and has suffered enough vaporization of ices to expose regions of dark surface material), then the shift from viewing a bright polar ice cap to a less reflective equator could explain why Pluto is dimmer than expected.

The orientation of Charon's orbit also alerted astronomers to a rare opportunity that was about to open whereby Pluto and Charon would reveal details on their surfaces. It was a priceless accident of timing. Pluto is so small and distant from Earth that, even when it is closest to the Sun, it is scarcely more than 0.1 arc second across, the size of a baseball seen from 100 miles (160 kilometers) away—too small for its disk to be measured directly from the surface of the Earth through the turbulence of our atmosphere. And Charon is smaller still. How is it possible to chart any detail at all on two worlds when their disks are not discernible?

The opportunity lay in a series of eclipses. When Charon was discovered in 1978, the south pole of Pluto was pointed forward in its direction of travel and a little bit sunward, just enough so that Charon circled Pluto without passing across the surface of Pluto or passing behind Pluto as seen from Earth. But Pluto's poles were pointed toward specific points in the star field. So, as Pluto slowly revolved around the Sun, observers on Earth would look directly first at one pole of Pluto, then its equator, then the other pole, then its equator again—over a period of 248 years. The orientation of Pluto and the orbit of Charon indicated that within a few years Pluto would arrive at a position in its orbit where, as seen from Earth, the orbit of Charon would be nearly edge on to Earth and Charon would be transiting across the face of Pluto and then vanishing behind Pluto in the course of its 6.39-day orbital journey. Those eclipses, watched for by astronomers since 1979, were finally detected in 1985. The results were immediate and dramatic.

Shadow Dance

Eclipse season for Pluto and Charon. As Pluto moves around the Sun, we see Charon's orbit around Pluto from gradually changing angles. Twice in every 248-year revolution of Pluto, the orbit of Charon is edge on to Earth. During this 5-year period, Pluto and Charon take turns eclipsing one another. In the course of each 6.4-day revolution of Charon around Pluto, first Charon passes in front of Pluto, blocking part of it from view, then Charon moves behind Pluto and is oc-culted. Because of the information about Pluto and Charon that these eclipses provide, it is very fortunate that Charon was discovered in 1978 rather than in 1990.

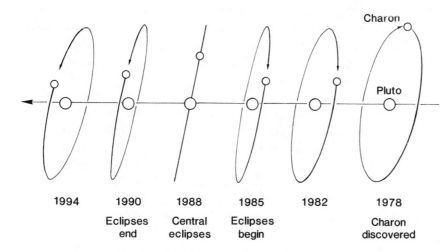

1994	1990	1988	1985	1982	1978
	Eclipses end	Central eclipses	Eclipses begin		Charon discovered

Astronomers carefully measured the amount of light coming from the Pluto-Charon system. The two are brightest when they are side by side to the line of sight—neither blocking the light of nor casting a shadow on the other. When Charon is in transit across the face of Pluto, the combined light of the planet-moon system declines by about 40 percent (about 0.55 magnitude) as Charon blocks part of the reflective surface of Pluto from view. As Charon emerges from transit, the brightness of Pluto-Charon is again at a maximum. Then Charon ducks behind Pluto and the brightness declines again, but only by about 20 percent (about 0.24 magnitude).

The occultation of Charon by Pluto does not subtract from the system as much reflected light as when Charon stands in front of Pluto and blocks a portion from view. Thus Charon must be darker and less reflective than Pluto. Pluto reflects perhaps 46 percent of the light it receives from the Sun; Charon only about 34 percent.

But the gleanings from Pluto-Charon eclipses are far greater than an approximation of comparative reflectivity. Previously, no markings on the surface of Pluto could be resolved by telescope. The only features on Pluto known to exist were the undefined surface oddities that caused Pluto's brightness to vary over its rotation cycle of 6.39 days.

In 1984, Marc W. Buie and Robert Marcialis, working independently, developed computer models of surface features on Pluto that could closely approximate the fluctuations in the planet's 6.39-day light period. Using the discoveries of methane ice on Pluto and the dimming effect as Pluto's equator was oriented more and more toward Earth, both scientists envisioned Pluto with extensive polar caps and a darker equatorial belt. Both also hypothesized that Pluto had two large equatorial spots. Marcialis used two dark spots in his model; Buie used one very large dark spot and a smaller bright spot. Which, if either, was correct?

In the series of eclipses between Charon and Pluto, each transit of Charon across the face of Pluto blocked different terrain from view. When the eclipse season began in 1985, our line of sight caused us to see Charon moving across the north polar region of Pluto. In 1988, midway through the eclipse season, Charon was transiting along Pluto's equator. As the eclipse season ends in 1990, our view will show Charon cutting across Pluto's south polar region. In the course of the eclipse cycle, Charon will have systematically blocked from view the entire Charon-facing hemisphere of Pluto. Gradually emerging as the eclipses proceed is a rough map of Pluto's surface in terms of light and dark features. So far, the gradations in tone favor the Buie two-spot model.

The smaller of the two spots in Buie's model—perhaps 250 miles (400 kilometers) in diameter—is as reflective as the polar caps. The larger of

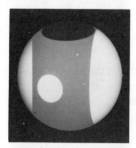

Pluto is so tiny and distant that telescopes on Earth cannot distinguish its disk; thus no surface features may be discerned on it. But based on light variations as Pluto spins on its axis in 6.4 days and as it revolves around the Sun, Marc W. Buie created a computer model of the surface of Pluto that shows bright polar ice caps, a darker equatorial belt, and two equatorial spots: one bright and the other very large and dark. Here Buie shows the rotation of Pluto bringing the light and dark spots into view. Pluto lies on its side as it revolves around the Sun. Its south pole is to the left. The hemisphere of Pluto with no spots is the one that always faces toward Charon. The present series of eclipses between Pluto and Charon should determine whether Buie's model is correct.
Courtesy of Marc W. Buie, UH/SDSC

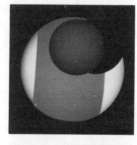

Using his computer model of Pluto, Marc W. Buie illustrates the March 19, 1987, transit of Charon across the face of Pluto. Pluto's south pole is to the left. Notice how large Charon is compared to Pluto and that Charon is darker in color. The amount by which the transit of Charon and its shadow reduce the light from Pluto tells astronomers how light or dark the eclipsed surface of Pluto is and allows a rough map to be drawn. The transit event shown here required approximately four hours.
Courtesy of Marc W. Buie, UH/SDSC

the spots is quite dark and has a width of 500 miles (800 kilometers), more than a third the diameter of Pluto. The bright polar caps and equatorial bright spot are interpreted as methane ice. The less reflective equatorial band may be a region partially covered with methane ice, with some exposed dark surface. The large dark equatorial spot might be nearly ice-free terrain where the dark pitted surface of Pluto is most visible.[14]

A Tour of Pluto

If a spacecraft from Earth (call it Orpheus) one day plunges into the realm of Pluto, what will its instruments—or its passengers—see?

Here is some speculation.

We have landed at the equator. The Sun shines brilliantly, far outshining every star in the sky. But at Pluto we notice a difference. Daylight is dim—like the dark overcast of a thunderstorm on Earth. The Sun is so distant that it is no longer a golden orb but now just a dazzlingly bright star in the heavens. For the first time, at the distance of Neptune and Pluto, we can see the Sun for what it is—a star.[15] We are so far away that the Sun is scarcely more than a point of light. Yet the other stars are still so distant that the Sun outshines the brightest of them by a factor of 10 million. After all, if we were on our way from the Sun to the nearest star, reaching Pluto would be like a runner completing the first 15 feet (4.7 meters) of a marathon. The comparison between the Sun and stars can be made directly because the stars can be seen in the sky at the same time as the Sun. Pluto's atmosphere is so thin that there is almost no scattered light to obscure the stars.

Here, with Pluto as close to the Sun as it can get, the Sun is about 900 times fainter than it appears on Earth. Still, the Sun shines 400 times more brightly on Pluto than the full Moon shines on Earth.[16] We can read a book in the sunlight on Pluto. But we must be careful not to cast a shadow on the page. The atmosphere on Pluto is so minimal that there are too few gas molecules and too little dust to scatter the sunlight and keep shadows from being absolutely black. What the sunlight falls on, we can see. What is in shadow disappears, unless Charon is visible and in a full enough phase to provide shadow-illuminating light. Contrasts on Pluto are very stark, like the Apollo astronauts found on the Moon. A rule for visitors to Pluto: Don't step into shadows. It would be easy to fall into a crater we could not see.

On our approach for landing on Pluto, we saw both of the polar ice caps, extending well toward the equator. Infrared measurements from Earth indicated that Pluto was partially covered by frozen methane, and much of that methane ice now lies at the poles, where the sunlight is now falling most obliquely and warming the highly volatile methane least well. It was not possible to tell during our approach whether the frozen methane was a deep ice field or a thin coating of methane frost.

With Pluto near perihelion, temperatures on the planet are near a maximum where the Sun is highest in the sky, which just happens to be at the equator. The temperatures are not exactly balmy—about 370° below zero Fahrenheit (−223° Celsius).

We have chosen to land at the equator because there is less frozen methane here, so that the surface of Pluto lies exposed. It is a sharp contrast to the bright white polar caps. The equatorial region is, on the whole, two to three times less reflective, with tones varying over large and small areas. Our overall impression of color is neutral medium gray, much lighter than the charcoal gray soil on Earth's Moon, but far off white. As we look harder, some, maybe much, of Pluto's gray surface seems to be just a tinge reddish brown, suggesting that the soil contains significant quantities of carbon and carbon compounds.

Pluto is certainly a battered world. Craters and debris are everywhere. Mapping from orbit should tell us whether the planet is saturated with craters and has changed little over the 4.6 billion years since the solar system began or whether it has erased some or many of its scars by surface upheavals due to internal processes rather than passively enduring a steady pounding from passing comets. If Pluto is little changed since the formation of the solar system, it could tell us much about the cloud of gas and dust from which the Sun and planets formed.

Craters of all sizes dominate the landscape. The rubble littering the barren ground everywhere is debris from eons of impacts. The horizon is hilly and uneven, with a suggestion of a couple of small flat-topped mountains projecting over the horizon—probably the rims of large craters. It may be that our landing site is within an ancient large crater—it is hard to tell. How well named Pluto is. The desolate frozen landscape, the dim and eerie light—it is not hard to imagine that the god Pluto lives here.

As we begin to walk about, we feel very light on our feet, and with each step we take, we bounce high above the ground. We have seen videotapes of the Apollo astronauts bounding across the Moon in one-sixth Earth gravity. Pluto's surface gravity is even weaker: less than one-third of our Moon's and only 5 percent what we experience on Earth. Here on Pluto, a 160-pound (73-kilogram) man weighs 8 pounds (3.6 kilograms). On Earth, from a standing start, we could jump about a foot (0.3 meter) into the air. With a flexible lightweight spacesuit on Pluto, we could jump about 20 times as high. Garage jumping could become a popular sport.

We will have to be careful not to venture too far from our lander without a homing device. Distance is hard to judge on a world where there are no familiar objects like houses to give a sense of scale and no haze to make the horizon look distant. Besides, on a small world like Pluto, the horizon not only looks sharp and nearby, it is truly very close. If we wander just 1.75 miles (2.75 kilometers) from our spacecraft, an astronaut standing on the ground there would lose sight of us below the horizon.[17]

One of our crucial experiments here will be to drill a hole in Pluto and pull out a core sample to see what lies below the surface. The abundance of water elsewhere in the outer solar system, including Charon, and the modest density of Pluto point to water being a major component of the planet, although its presence on Pluto has not yet been confirmed because the spectrum of methane overwhelms the spectral bands of water. Does frozen water lie trapped below the surface? Perhaps much of the terrain upon which we are standing is water ice mixed with flecks of carbon and other compounds, which gives it a darker cast.

Over there—and, yes, elsewhere, now that we look closely—are white frosty patches. We bounce over for a closer look. (It's hard to stop bouncing.) It is a thin layer of methane frost in beautiful large crystals. The sublimation of methane from ice to gas provides Pluto with a very tenuous atmosphere. Some of the methane vaporizing here will drift toward the poles and freeze on the ground there until the changing seasons bring summer to the north pole in 2051. But by then, Pluto will be much farther from the Sun and less methane will sublimate. Pluto's poles stay covered with methane ice throughout the planet's 248-year period of revolution.

On our descent for landing, flying along Pluto's equator, we saw (in our continuing speculation) a large white feature, perhaps 250 miles (400 kilometers) across. It looked like a refugee from the polar ice caps. Yet we knew to watch for it. It was predicted by Buie's two-spot model of Pluto, based on light variations as the planet rotates. The roughly circular shape of this intensely white spot suggests an impact feature. Has it filled with ice welling up from below, like that bright crater on Uranus' Umbriel? Is it a young feature that has not yet had time to darken from exposure to sunlight and fast-traveling subatomic particles from the Sun and stars? Why it should be so different from the surrounding terrain is not obvious.

We hope to explore that equatorial bright spot later, but we have deliberately landed near an even larger feature and are making our way there now. On light/dark maps of Pluto it appeared as an extremely dark spot perhaps 500 miles (800 kilometers) across—more than a third the diameter of Pluto.

Computer models of Pluto's surface used circular spots for convenience to approximate the sizes and positions of features that could provide the planet with the light variations it shows. Does an actual feature exist where the model shows a huge dark circular spot? A giant impact basin has this shape. Imagine it, stretched out before us: an enormous crater, its walls and floors pocked by other craters. We are surrounded by desolation. This crater is so immense that it quickly stretches out of sight beyond the horizon in front of us and to our sides. It must be primarily this dark feature that created the light variations from Pluto that first allowed the length of its day to be measured at 6.4 Earth days. But if this basin was formed by impact, how could Pluto have survived the shock? At the very least, the impact would have caused such upheavals within Pluto that the scar would have been erased.

Whatever surface features may correspond to the small bright spot and large dark spot in Buie's model, they are located on the side of Pluto always facing away from Charon. We return to the side of Pluto where Charon can always and only be seen—a hemisphere with only slight contrasts.

Moonlight

We look up at Charon in the sky, 20 times closer to Pluto than our Moon is to Earth. It is an impressive sight. Charon may rank twelfth in size among moons in the solar system, but it is so close to Pluto—only 11,650 miles (18,800 kilometers) above Pluto's equator—that it appears larger than any other moon appears from the surface of its planet. Charon covers almost 4 degrees in Pluto's sky—eight times as wide as our Moon appears from Earth. On our planet, you can hold a pea out at arm's length and completely eclipse our Moon. On Pluto, to block Charon from view, you would need a billiard ball.

It was no surprise that Charon rotates in the same period of time as it revolves so that it always presents the same hemisphere to Pluto. All the inner satellites and all the major satellites in the solar system have synchronous rotation and revolution because they are tidally coupled to their planets. A planet's gravity creates a slight tidal bulge in its moons and pulls on that bulge so that the moons cannot turn it away from the planet. One side of the satellite always faces the planet and the other side always faces away while the planet rotates rapidly, so that the moon rises and sets for all parts of the planet.

But Pluto furnished a surprise. Pluto and Charon are so close to twins in size and so close together that Charon's gravity induces a bulge in Pluto. The bulge is great enough that Pluto is tidally coupled to Charon just as Charon is tidally coupled to Pluto. Thus, Pluto always shows the same face to Charon just as Charon always shows the same face to Pluto. It is the only example of mutual tidal coupling in the solar system.[18] The result is that for an astronaut standing on Pluto, Charon is either always visible or never visible. If it is visible from that location, Charon never rises or sets and always stands at the same place above the horizon. If

NO TIDAL COUPLING (Not to scale)

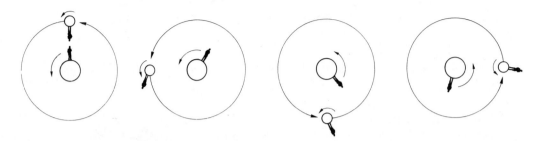

SATELLITE TIDALLY COUPLED: EARTH/MOON SYSTEM
(Moon shows only one side to Earth)

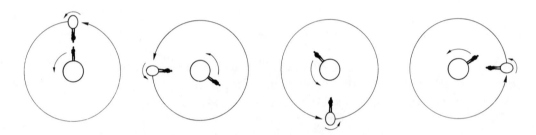

PLANET AND SATELLITE BOTH TIDALLY COUPLED: PLUTO/CHARON SYSTEM
(Both bodies show only one side to each other)

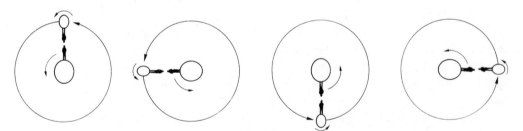

**Planet-moon systems
with and without tidal
coupling**

Charon is not visible from the astronaut's landing site, it will never rise, and the astronaut will have to make a pilgrimage to the proper side of Pluto to enjoy moonlight.

We have deliberately landed on the hemisphere of Pluto where Charon is always seen. Charon far outshines all other bodies in the sky except the Sun. And it is close enough to show surface features even without a telescope. The shadows we see on Charon reveal an uneven, cratered landscape. Like Pluto, Charon is light gray, although somewhat darker and more even in color than Pluto, as was known from measurements made from Earth using the Pluto-Charon eclipses. The very slightly reddish brown hue of Pluto is missing from Charon—or at least from Charon's Pluto-facing side, that is the only side we get to see from the surface of Pluto. Missing too from Charon is the methane frost which partially covers Pluto. With Charon's smaller mass and therefore weaker gravity, whatever methane ice there was at the surface has evaporated.[19] Perhaps this in part explains why Charon is less reflective. The escaping methane has exposed frozen water to view.

On Earth, we are used to the rising and setting of the Sun, Moon, and stars as our planet turns. On Pluto, the Sun rises and sets, if somewhat slowly, but Charon stays fixed in the sky.[20] It never rises or sets, thanks to tidal coupling. As Charon revolves once around Pluto in 6.4 days, Pluto spins once around on its axis in that same period of time. The result is that Charon hangs almost stationary in the sky while the Sun and stars glide slowly past in the background. Because Charon is so large in the sky, stars are frequently blocked from view. These stellar occultations are the only eclipses visible during the 120-year gap between seasons of solar and lunar eclipses.

From the vantage point of Earth, Pluto and Charon pass in front of and behind one another very rarely. The Earth experiences solar and lunar eclipses at least four times and sometimes as many as seven times a year. Because of Pluto's axial tilt and Charon's position over Pluto's equator, the pair go for almost 120 years without their shadows ever falling upon one another. Then, in a period roughly six years long, Charon's orbit is nearly edge on to Earth and every 6.39-day orbit Charon makes carries it across the face of Pluto and then around behind Pluto. The result is an eclipse frenzy. Serendipitously, that eclipse season began in 1985, soon after Charon was discovered and while Pluto and Charon were near perihelion (hence easiest to study). The eclipse season will continue into October 1990. After 1990 there will be no more occultations or transits of the pair until about 2109, and then Pluto will be its farthest from the Sun and its least accommodating for study from Earth.

For an astronaut on Pluto in 1988, at the peak of the present eclipse season, a solar eclipse is a stirring sight. The Sun is by far the brightest object in the sky, but in apparent size, it seems to be no more than a starlike dot. The rotation of Pluto carries this dazzling point of light closer and closer to the huge disk of Charon until it disappears—dimming out at Charon's limb in a little less than 30 seconds.[21] With the Sun passing behind the widest part of Charon's disk, the total phase of this solar eclipse would last 1 hour 37 minutes. (The total phase of solar eclipses on Earth last a maximum of 7 minutes.)

During an eclipse of the Sun on Pluto, Charon would look like a giant dark hole in the sky, marked only by the absence of stars. It would be dark but not black because it would be illuminated by reflected light from Pluto. The corona—the outer atmosphere of the Sun, which makes solar eclipses seen from Earth so beautiful—would be visible only just after the Sun vanished and just before it reappeared. At mid-eclipse, the disk of Charon covers the entire orbit of the Earth. The corona is far too faint at that distance from the Sun to peer around the edges of Charon.

The Origin of Pluto

How could this strange double-planet system have developed? The rest of the planets in our solar system lie in a plane that is so wide and yet so flat that if it were shrunk to a size you could hold in your hand, the solar system out through Neptune would be about the size and thickness of a *Time* magazine. But Pluto violates that analogy.

Of all the planets, the inclination (17.2 degrees) and the eccentricity (0.25) of Pluto's orbit is the greatest. Its orbit is so elliptical that Pluto sails out to a distance of 49.3 astronomical units—almost 4.6 billion miles (7.4 billion kilometers) from the Sun. Since its last aphelion in 1866, the year after the conclusion of the Civil War, the elliptical orbit of Pluto has been carrying it ever closer to the Sun. On January 21, 1979, Pluto passed inside the orbit of Neptune—the only planet that crosses the orbit of another. Pluto will spend 20 years closer to the Sun than Neptune and will not resume its usual distinction as the outermost planet until May 14, 1999. When *Voyager 2* reaches Neptune on August 24, 1989,[22] it will literally be visiting the farthermost planet from the Sun.

The modest size of Pluto that was emerging after its discovery and its excursion within the orbit of Neptune suggested to some astronomers that Pluto's ancestry could be traced back to Neptune.

In 1934, Issei Yamamoto of Japan proposed in a lecture that Pluto was an escaped moon of Neptune. Yamamoto suggested that a star passed close to Neptune, causing Neptune to fall to an orbit closer to the Sun. The encounter also forced Triton into a retrograde orbit and cast Pluto out onto a planetary orbit centered on Neptune's original distance from the Sun (in agreement with Bode's Law for an eighth planet).[23]

In 1936, Raymond A. Lyttleton of England attempted to explain Pluto's curious orbit by suggesting that Pluto had originally been a moon of Neptune but that a near collision between Triton and Pluto had gravitationally accelerated Pluto out of the Neptunian system and had radically altered the course of Triton so that it now moved around Neptune in the opposite direction—the largest satellite with retrograde motion and the only satellite with that orbital anomaly that lies close to a planet.

In 1956, Gerard P. Kuiper agreed that Pluto had come from the Neptunian system, escaping as the planets completed their formation. But Kuiper saw no need for a near collision with Triton to eject Pluto. Instead he thought that Pluto had abandoned Neptune when the Sun began to shine. The Sun's radiation and particles had, he said, swept away the light gases at Neptune, reducing the planet's mass by a factor of 40. Neptune then had too little mass to retain Pluto by gravity.[24]

Many astronomers, however, were uncomfortable with the idea that Pluto had come from the Neptune system. They pointed out that when Pluto crosses the path of Neptune, it is always 8.9 astronomical units north of Neptune's orbit on its way inbound to perihelion and 6.1 astronomical units north of Neptune's orbit outbound. At its closest, Pluto is about as close to Neptune's orbit as the Earth is to Jupiter when they are on opposite sides of the Sun. The orbits of Neptune and Pluto never intersect like a road at a railway crossing. Instead the passage is more like a bridge high above the railroad track. When it moved temporarily inside the orbit of Neptune in 1979, Pluto passed far above the path of Neptune, and Neptune was nowhere near that point in its orbit.

If Pluto crosses above Neptune's orbit at a minimum distance of 6.1 astronomical units, it would seem that every once in a long while Pluto and Neptune must come as close to one another as 6.1 astronomical units. But according to careful mathematical analysis and computer simulation, Pluto and Neptune can never get closer to one another than 16.7 astronomical units—almost the distance from the Earth to Uranus.[25]

Picture Pluto's orbit as a moderately elongated ellipse made of stiff wire. Normally, a planet revolves around its orbit like a bead strung on the wire while the entire orbit—the stiff wire—very slowly rotates (precesses) like a very fat hour hand on a clock.

The reason that Neptune and Pluto never come close together is that the size of the gravitational forces that Neptune, Uranus, Saturn, and Jupiter exert on Pluto and the timing of those forces do not allow Pluto's orbit to precess but only to oscillate within a very narrow range. This confinement appears to be permanent, so Pluto's and Neptune's orbits can never and have never come close to actually intersecting so that a collision or close encounter would be possible. Further, the rhythmic gravitational effects of the outer planets slightly adjust the size and shape of Pluto's orbit, and thus its period of revolution, so that Neptune is never nearby when Pluto crosses over Neptune's orbit.

Curiously, Neptune and Pluto are closest to one another when Pluto is near aphelion—about as far beyond the orbit of Neptune as it can get.

Pluto can actually come closer to Uranus (10.6 astronomical units) than it can to Neptune.

Recently, astronomers have used supercomputers to calculate the paths of Pluto and Neptune hundreds of millions of years backward into the past and forward into the future, trying to take into account the gravitational perturbations they caused in each other's orbit and the perturbations that the other giant planets caused in theirs. These studies have found no incident where Neptune or Pluto were precariously close to one another.[26]

The discovery of Charon greatly changed the perspective on Pluto's origins, just as it changed many perceptions of Pluto. First, the motion of Charon showed that Pluto was even smaller than previously imagined. There was no way that Pluto, as a moon of Neptune, could have reversed Triton's orbital direction. With such a low mass, if Pluto, as a moon of Neptune, had ever received an acceleration from a close encounter with

IS PLUTO A PLANET?

As information about Pluto has increased, the size attributed to Pluto has decreased. It is certainly the smallest planet, far smaller than Mercury. It is considerably smaller in size and mass than our Moon. At the same time, Pluto has shrunk in the estimation of some astronomers to the point that they are calling for the expulsion of Pluto from the ranks of planethood and its demotion to the category of asteroid.

How does Pluto fit in with the minor planets? Lousy. First of all, it's too large. It is more than twice the diameter of the largest asteroid, Ceres. Even Charon, Pluto's moon, is larger than Ceres. In fact, Pluto has about three times more mass than all the asteroids in the solar system put together.

Second, Pluto is in the wrong place to be an asteroid. The vast majority of minor planets orbit the Sun between Mars and Jupiter. The only known minor planet beyond Jupiter that doesn't return to this asteroid belt for part of its orbit is Chiron, whose circuit carries it from just inside the orbit of Saturn to just inside the orbit of Uranus. It never gets within a billion miles (1.6 billion kilometers) of Pluto. Chiron's mean distance from the Sun (13.7

astronomical units) places it closer to the asteroid belt than to Pluto.

If a second asteroid belt is discovered beyond Neptune and if Pluto is fairly typical of the size and composition of those minor planets, then it might make sense to recategorize Pluto as an outer asteroid. But Clyde Tombaugh and, 40 years later, Charles Kowal diligently searched the heavens for Pluto-sized objects beyond Neptune. They found none, and it is most likely that objects even close to that size do not exist or at least are very rare.

Pluto is different from the known asteroids in other ways as well. Its much greater mass allows it to retain a moon and a tenuous atmosphere.

Pluto just doesn't qualify as an asteroid. We learn little or nothing about the structure of the solar system by approaching Pluto as a minor planet.

How then does Pluto fare as a comet? Better, but still poor. Pluto's distance from the Sun places it not too far inward from the possible inner edge of the Oort Cloud of comets. Pluto's composition and the ellipticity and inclination of its orbit are somewhat cometlike. But again, Pluto is far too large. The nucleus of a typical comet is perhaps 5 to 10 miles its master, Pluto would have been ejected from the solar system, not just the Neptunian family.

Since gravitational interaction within the Neptune system between Triton and Pluto would not have been powerful enough to have caused the peculiar orbits of Neptune's satellites, Robert S. Harrington and Thomas C. Van Flandern in 1979 proposed a different source of energy. They suggested that Triton, Pluto, and Charon were indeed once moons of Neptune but that their orbits had been disrupted by an outside intruder—an undiscovered tenth planet with two to five times the mass of Earth. Pluto and Charon were ejected from the Neptune system close enough together to be gravitationally bound.

But most astronomers now believe that Pluto and Triton are planetesimals left over from the beginning of our solar system. Other planetesimals that reached their size merged together to form still-larger bodies such as Uranus and Neptune. Pluto (with its satellite Charon) was

---

(8 to 15 kilometers) across. Pluto is about 1,457 miles (2,345 kilometers) in diameter.

Pluto is also somewhat too dense. Comets are dirty snowballs with a density close to that of water. Pluto's density is twice as great, indicating that it is about 70 percent rock and 30 percent ice. Its density is similar to Triton's and Neptune's and intermediate between those of comets and asteroids.

Pluto—as well as Uranus, Neptune, and their icy satellites—probably originated from the accretion of comets, just as Earth and the other rocky midget planets near the Sun probably formed from rocky planetesimals of which the asteroid belt is the least altered remnant. But no one considers the Earth an asteroid even though its progenitors were asteroid-like. So, likewise, it would be unhelpful to classify Pluto as a comet.

There is simply nothing to be gained by excluding Pluto from planethood. There is no asteroid belt known beyond Neptune for which Pluto can serve as a charter member. Even if Pluto is an overgrown comet, it is far too large to be representative of that species. Therefore, to designate Pluto as an asteroid or comet is to make it an orphan—too different from its brethren for useful comparison. To desig-

nate Pluto an asteroid or a comet is to ensure that it will virtually drop from treatment in classrooms and schoolbooks, and a new generation of children will grow up thinking of Pluto as a "mistake," if they have any awareness of Pluto at all.

And then think how easy it will be to convince any Congress that NASA deserves funds to send a spacecraft to an extremely distant asteroid.

No. Pluto is a planet—by orbit and by size. It is best understood—and taught in astronomy courses—as a probable example of a large outer solar system planetesimal, the kind of object formed from cometlike bodies in the outer planetary realm at the beginning of our solar system. Most of these planetesimals merged with others to mold the comet conglomerates we now call Uranus, Neptune, and their moons.

But there may be a pair of planetesimals that did not coalesce further. Pluto and Charon have potential value to science not as representatives of a class of asteroids or comets but as large, nearly pristine planetesimals—the objects from which gas giant planets and their moons are made.

locked by the gravity of the outer planets in a resonant orbit that protected it from collision with and accretion into a larger planet. William
B. McKinnon has suggested that Triton survived until planet formation
was nearly complete but then passed close by Neptune, was slowed by
the gas and dust remaining around the planet, and was captured by Neptune's powerful gravitational field into a retrograde orbit.

Such a capture would have caused intense heating inside Triton, melting
the interior and allowing the newly acquired satellite to differentiate.
If so, the surface of Triton should still attest to its stressful early life. Pictures and measurements from *Voyager 2* at Triton may indicate whether
it was indeed captured by Neptune.[27]

Although Pluto never experienced the stresses imposed on Triton by
Neptune, its interior too has probably undergone major changes since
accretion. The presence of low-density methane ice everywhere on the
surface of Pluto and yet the substantial density of the planet, about twice
that of water, suggests to McKinnon and Steve Mueller that Pluto's interior must have melted and separated into layers.

Pluto's density indicates that it is composed of much more rock than

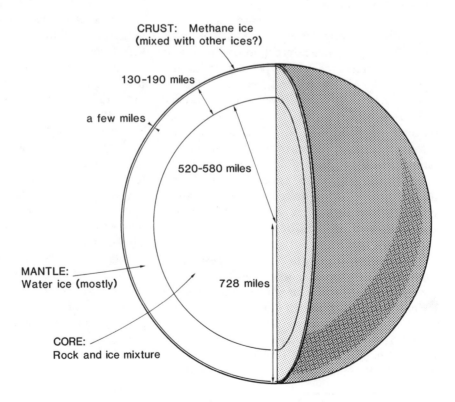

CRUST: Methane ice
(mixed with other ices?)

130-190 miles

a few miles

520-580 miles

MANTLE:
Water ice (mostly)

728 miles

CORE:
Rock and ice mixture

**Internal structure of
Pluto as proposed by
William B. McKinnon
and Steve Mueller.**

ice. The quantity of rock is great enough so that the decay of radioactive elements soon after the planet formed must have caused Pluto's interior to differentiate, with the denser rocky materials sinking toward the center and the lighter ices floating toward the surface. With methane ice concentrated toward the surface, slow vaporization could have kept Pluto supplied with a thin methane atmosphere since the solar system began.[28]

McKinnon and Mueller also suggest that Charon was once part of Pluto. A large planetesimal wandering through the outer solar system smashed into the original, larger Pluto. But the orbital velocities of objects so far from the Sun are not so great that the collision would have scattered the debris widely. Instead, the fragments of the original Pluto, continuing to orbit together, could have reaccreted but into two smaller bodies orbiting one another—Pluto and Charon.

As perhaps the only remaining large planetesimals left in the outer solar system, Triton and Pluto-Charon could be closely related. Since Pluto will soon be gliding farther from the Sun again and since no spacecraft to visit it has yet been planned, it may be that Triton is our best clue to the nature of Pluto, unless Triton's capture by Neptune has changed it too greatly to provide useful comparisons.

A Mission to Pluto?

On September 12, 1989, Pluto will celebrate perihelion, its closest approach to the Sun, when it reaches a distance of 2.75 billion miles (4.4 billion kilometers). Astronomers—professional and amateur—will join the celebration because Pluto at perihelion is seven times brighter than at aphelion. When it is farthest from the Sun, Pluto sinks to magnitude 15.9 and a 22-inch (56-centimeter) telescope is about the minimum size required to study the distant planet. At perihelion Pluto is expected to reach a maximum magnitude of 13.6, so an 8-inch (20-centimeter) telescope should just be able to discern it.[29]

During 1999, Pluto will return to its role as the outermost planet, gradually sliding back into the darkness of distance that helped to hide it from discovery and then from understanding for so many years. The next decade or two offers a rare window of opportunity to study what for 228 out of its 248-year orbit is the most remote planet in our solar system.

Will there be a space probe to visit Pluto and Charon in the near future? NASA has no plans, but the Soviet, European, Japanese, and Chinese space programs are expanding. Now is the time for a mission to Pluto, while the planet is still about as close to the Sun as it ever gets. Send a spacecraft to Pluto soon and save 1 billion miles (1.6 billion kilometers) over the average distance from the Earth to Pluto. That's a savings of greater than the distance between the Earth and Saturn.

A direct flight from the Earth to Pluto using a least-energy orbit would take 40 years. But by using Jupiter for a gravity assist, the time for the journey would be cut to 10 to 15 years, and far less energy would be required. The next launch window for a Jupiter-assist flight to Pluto is in the late 1980s—but no spacecraft has been built. The next opportunity will come in 12 years when Jupiter completes another revolution of the Sun and begins to overtake Pluto once more. Perhaps there will be an orbiter and lander mission on its way to the outermost planet and its outsized moon in the year 2001.

CHAPTER 13

Is There a Tenth Planet?

"The finding of Pluto was an important discovery, but what you did not find out there is even more important."
Astronomer Gerard P. Kuiper to Clyde Tombaugh
(1950)

As soon as Tombaugh had discovered Pluto and reconfirmed it with new photographs, he resumed his systematic photographic survey in Cancer and Leo where he had left off. This self-assigned use of time was probably a good release of nervous energy because almost four weeks would pass between the moment he delightedly called across the hall to Lampland and the actual announcement of a ninth planet. However, said Tombaugh, "I suspended blinking, thinking the search was over—so I thought. I was getting tired of the tedious blinking anyway."[1]

After ten weeks of excitement, the pressure for information from other observatories and from journalists had subsided. One day in late May 1930, Vesto Slipher went to Tombaugh and made a request: "I would like you to resume searching. There might be more planets out there like Pluto. Start where you left off in February." Tombaugh was pleased: "The staff now had complete confidence in my work. I had examined two million stars. To cover the wider area, I had at least twenty million stars to go." That number proved to be an underestimate.

He steadily expanded his search north and south of the ecliptic to take in all the sky visible from Flagstaff. He kept a special eye out for the several planets that William H. Pickering had predicted. "As I was to learn over the next thirteen years in the resumed search," wrote Tombaugh, "not one of his other planets existed."

Tombaugh worked on resolutely: "After the experience with greed involved in the Pluto discovery, I could see that even in astronomy, it was a dog-eat-dog situation. After all the sacrifices made by the Lowell family and the staff, I was determined that if there were more planets to be found, they would be found at the Lowell Observatory."

By April of 1938, Tombaugh had photographed with the 13-inch telescope and examined with the blink microscope 35 million stars. By July of 1943 he had photographed the entire sky visible from Flagstaff, from Canopus to Polaris. But his tolerance for blinking was steadily dropping. Once he could maintain concentration for five to six hours per day. A few years after the discovery of Pluto, he had to cut back to three to four hours. Now he could endure only two hours a day. "I was burning out," he said.

In July 1943, in the midst of World War II, Clyde Tombaugh, now 37 years old, was drafted to teach navigation for the Navy in Flagstaff. With that event, said Tombaugh, "my planet searching ended forever."

His final tally of individual stars photographed and blinked numbered 45 million—90 million paired images. He had spent 7,000 hours at the blink microscope, the equivalent of doing nothing else for an eight-hour working day for three and a half years.

In his 14 years at the blink comparator, Tombaugh had found:

1 new globular star cluster
5 new open star clusters
1 new supercluster of 1,800 galaxies
several new small clusters of galaxies
1 new comet
about 775 new asteroids
and
1 new planet

Because of his careful and dogged years of search, Tombaugh was confident that no planet brighter than magnitude 16.5 existed within his field of coverage. Only a planet in an almost polar orbit and situated near the south pole (not visible from Flagstaff) could have escaped his detection. A planet the size of Jupiter would have been detected at 470 astronomical units (12 times Pluto's distance from the Sun). He could have picked up a Neptune-size planet at 7 times the distance of Neptune. He could have seen a planet the size of Pluto at 1.5 times Pluto's average distance from the Sun. "Other planets like Pluto do not appear to exist out to a distance of 60 astronomical units," Tombaugh concluded,

Beyond 60 astronomical units, said Tombaugh, more Pluto-like planets could exist, but their perturbations, like Pluto's, would be immeasurable. The only means to find them would be by a comprehensive observational search of lengthy, tedious, and costly proportions.

But if Pluto wasn't massive enough to have caused the irregularities in the motion of Uranus and (according to William H. Pickering) the motion of Neptune as well, what *was* causing those discrepancies? Had Planet X been missed?

As it gradually became apparent that Pluto was too small to be the culprit, predictions for the existence of a tenth planet began to appear fairly steadily, most based on new analyses of the orbital discrepancies.[2]

Some of the scientists who think there is evidence that points toward a trans-Plutonian planet and who have been most active recently in pursuit of Planet 10 are Thomas C. Van Flandern and Robert S. Harrington, Daniel P. Whitmire and John J. Matese, John D. Anderson, and Conley Powell. Each came to the conclusion that a tenth planet exists by a different route, but the planet each envisioned had much in common with the others.

The Quest Goes On

His work at the Almanac Office of the U.S. Naval Observatory kept Tom Van Flandern close to peculiarities in planetary orbits. The outer planets were uncooperative. The calculated orbits for Neptune would fit its observed position and provide useful predictions for a few years, but by the end of ten years, the predicted positions were considerably in error. A calculated orbit for Uranus would fit for one whole revolution of the planet but, maddeningly, would not fit the previous or a subsequent revolution. The orbits of some short-period comets also, he felt, showed the effects of perturbations that could not be explained by the gravity of known planets or by the jet effect of ice on rotating comets vaporizing so that it slightly alters the comets' orbits. Van Flandern became convinced about 1976 that the discrepancies in the motion of Uranus and Neptune were real and not just errors of measurement.

He then called the anomalies in the motions of Uranus, Neptune, and the comets to the attention of his Naval Observatory colleague Robert S. Harrington and suggested the idea of a tenth planet. Harrington was skeptical at first, but the discovery of a moon for Pluto showed that the mass of Pluto was far too small to disturb Uranus, Neptune, or distant comets.

Van Flandern and Harrington began their collaboration with a study of the peculiarities in the Neptune system. The newly calculated mass of Pluto had demolished Lyttleton's already seriously undermined idea that Pluto was once a satellite of Neptune along with Triton and that a gravitational encounter between Triton and Pluto had reversed the orbital motion of Triton and hurled Pluto out into its own orbit around the Sun. Pluto didn't have the mass to turn Triton around.

Could the gravitational energy to force Triton into retrograde revolu-

Piecing Scattered Clues Together

Thomas C. Van Flandern
Courtesy of Thomas C. Van Flandern

Robert S. Harrington
Official U.S. Naval Observatory photograph

tion have come from an outside source—the same source that they felt was disturbing the outer planets? Could that intruder have been a planet?

They made computer simulations of planets of different sizes intruding into the Neptunian system at different distances, speeds, and angles. From their computations emerged a nominee—a planet with two to five times the mass of Earth in a highly inclined and elliptical orbit 50 to 100 astronomical units from the Sun with an orbital period of about 800 years.[3] In one brief visit long ago, the intruder had reversed the motion of Triton, warped the orbit of Nereid, and cast the moon Pluto out of the Neptune family onto a planetary orbit of its own. Charon, they suspected, was either an additional satellite of Neptune expelled along with Pluto so that they captured one another; or, alternatively, the intensity of the intruding planet's tidal strain had caused Pluto to break in two, with Charon as the smaller fragment. For Van Flandern there was a third alternative: Charon might be a former satellite of the intruder planet that was transferred to Pluto's control as Pluto and the intruder sped away after the encounter.

With their inclined, eccentric path, Harrington and Van Flandern felt, Pluto and Charon hinted at their past by returning four times a millennium to a position near the scene of the crime—the place from which they were kidnapped and abandoned, the place of their birth.

The ideas of Van Flandern and Harrington then began to diverge. Van Flandern continued to think that the marauding planet had formed beyond Neptune at the beginning of the solar system. Harrington thought that Planet 10 had formed between the orbits of Uranus and Neptune.

As to commencing a search for Planet 10, Van Flandern felt it was too early. More data was needed, such as an improved determination of the mass of Neptune, which *Voyager 2* should furnish. Also needed was much more mathematical analysis to demonstrate that a tenth planet had to exist and where it could be found.

Harrington preferred to approach the problem of locating the suspected planet by brute force—supplying a computer with thousands of orbits to crunch in order to find those orbits that might be possible. He began searching for Planet 10 in October 1979, using the Naval Observatory's 15-inch (38-centimeter) astrograph (a wide-field photographic survey telescope) in Washington, D.C. In 1980, he continued, using the 24-inch (61-centimeter) Curtis Schmidt Telescope at the Cerro Tololo Inter-American Observatory in Chile. Additional searches were mounted in 1981 and 1984 with the 15-inch telescope in Washington, D.C., and then in 1986 and 1987 with the 8-inch (20-centimeter) double astrograph of the U.S. Naval Observatory's Black Birch site in New Zealand. He examined all the plates on a blink comparator and has, in his words, "nothing to show for my efforts."[4]

He and Van Flandern still agree that Planet 10 should be a frozen methane, ammonia, and water world somewhat like Uranus and Neptune but of lower mass—perhaps two to five times the mass of Earth. To remain so far unseen, Planet 10 must be nearing aphelion on its highly elliptical orbit, so that it is near minimum brightness. Still, for an icy body its size, this tenth planet should be about magnitude 13, some six times brighter than Pluto when it was found. Van Flandern suggests that a trans-Plutonian planet could be dark in color, reflecting light so poorly that it would be no brighter than 16th or 17th magnitude—fainter than Pluto at its discovery.

Harrington is not driven on in his quest by utter confidence in the irregularities in the motions of the outer planets. He thinks they are real but admits they are slight. Imagine, he says, observing from Washington, D.C., and identifying a drunk coming out of a Baltimore bar by his stagger. That's the size of the alleged perturbations in the motions of Uranus and Neptune. It may be, he feels, that these discrepancies are errors in converting older observations to a common reference frame. But he is encouraged to keep hunting by the computer simulation he and Van Flandern performed, that found that a single planet of modest mass passing only once through the Neptune system could account for the present motions of Triton, Nereid, and Pluto.

And if he finds Planet 10, what will he name it? "Humphrey," he says. The name was inspired by "Humphrey the Camel," a somewhat off-color song from the 1960s. But Humphrey isn't exactly a traditional Roman or Greek name. "Well," he says, "I considered Zorba." When searching for a tenth planet, Harrington feels, commitment is necessary, but it doesn't pay to take yourself too seriously.[5]

The Killer Planet

In 1985 two more researchers postulated a tenth planet, but they approached the issue from a different direction. Daniel P. Whitmire and John J. Matese, at the University of Southwestern Louisiana, had been active in the controversy about the cause of mass extinctions on Earth, such as the one about 65 million years ago that wiped out the dinosaurs.

In 1980, Luis W. Alvarez, Walter Alvarez, Frank Asaro, and Helen V. Michel had proposed that the impact of an object from space had thrown so much dust into the atmosphere that the Sun was blocked from view. It was black as night for months. In the dark and cold, plants died. Animals that depended on the plants died. Animals that depended on other animals died. More than half the species of plants and animals on Earth perished at that time. The Alvarezes, Asaro, and Michel felt that the assassin was an asteroid perhaps 6 miles (10 kilometers) in diameter.[6]

But the extinction that claimed the dinosaurs and so much of life 65

John J. Matese (left) and Daniel P. Whitmire
Courtesy of Daniel P. Whitmire and John J. Matese

million years ago was only one of many mass extinctions that the Earth has experienced in the last 250 million years. In 1984, David M. Raup and J. John Sepkoski, Jr., presented evidence that large-scale obliterations of life on our planet have occurred approximately every 28 million years.[7]

Two different teams of researchers immediately recognized that asteroid collisions with Earth are too infrequent to cause mass extinctions every 28 million years and that there was no mechanism to get asteroids to hit the Earth with regularity. But comets could, if something was disturbing them. They proposed that the periodic mass extinctions were triggered by an unseen star within our solar system, bound to our Sun by gravity, just as the planets are. After all, a majority of the stars in the universe are binary or multiple star systems.

One team of researchers was Marc Davis, Piet Hut, and Richard A. Muller; the other team was Daniel P. Whitmire and Albert A. Jackson IV. Their proposals were independent, simultaneous, and very similar. The Sun's companion had to be a star of low mass and hence low brightness to remain unidentified when it was by far the closest star to the Sun. This star had an elliptical orbit that carried it in as close as 0.5 light-year (about 30,000 astronomical units) to the Sun and then out to a distance of 2.4 light-years (more than halfway to the nearest star) in a period of 28 million years.

In the course of its travels, this star would at perihelion careen through the densest portion of the Oort Cloud of comets, accelerating millions of them out of the solar system but decelerating millions of others, forcing them to fall in closer to the Sun so that some hit the Earth with momentous consequences for the evolution of life. Davis, Hut, and Muller dubbed this small but lethal hypothetical star Nemesis, "after the Greek goddess who relentlessly persecutes the excessively rich, proud, and powerful. We worry," they added, "that if the companion is not found, this paper will be our nemesis."[8]

There is, however, no consensus among paleontologists and geologists about the causes of mass extinctions and whether mass extinctions and cratering episodes occur at regular intervals. The controversy continues at a lively pace into the present.

Were mass extinctions caused by comet showers triggered by a companion star for our Sun? One of the early attacks on this hypothesis came from analyses of the stability of a companion star on such an elliptical orbit that carried it more than half the distance to the nearest star. It was doubtful that it could survive in such an orbit from the beginning of the solar system because the Sun's gravity at that distance is very weak. Stars and clouds of interstellar gas and dust passing near the solar system would have disrupted the regularity of the twin star's period and, rather quickly, would have torn it away from the Sun.

In proposing a "death star" hypothesis, Whitmire had already recognized the problem of a companion star's orbital stability. He and John J. Matese sought an alternative: What if the culprit were a planet instead? Instead of Sun 2 scattering comets with a plunge from the fringes into the densest regions of the Oort Cloud, this Planet 10 would disrupt comet orbits by scratching the innermost portion of the Oort Cloud from inside, in particular a disk of comets thought to lie in the plane of the solar system with orbits not far beyond Neptune and Pluto.

Yet Pluto takes only 248 years to orbit the Sun. How could a planet not too far beyond Pluto take 28 million years to complete a circuit? According to the law of gravity, it couldn't. Even a planet several times the distance of Pluto cannot revolve around the Sun that slowly. But the entire orbit of a planet will revolve due to the gravity of neighboring planets, and this precession can proceed very slowly.

Visualize Planet X on an elliptical orbit perpendicular to the plane of the solar system. If the plane of the solar system is a table top, then the orbit of Planet X is standing on its head. We see the table top at eye level edge on and the orbit of Planet X face on. In that headstand position, Planet X does not disturb any disk of ancient comets that lie in the plane of the solar system not far beyond Neptune and Pluto. But eventually it will disturb the comets in the disk because the gravity of the outer planets causes the entire orbit of Planet X to precess. The tilt of the orbit will stay inclined to the plane of the solar system by, say, 90 degrees, but the orbit will gradually pivot around like a very fat hour hand on a clock. It will pivot through the plane of the table and down, then back up through the plane of the table and back to its initial headstand position. In a multiple-exposure photograph, the ellipses of Planet X's orbit would form a rosette.

In the course of that precession, the outer portion of the orbit (the end of the fat hour hand) would slide through the comet disk only at the nine and three o'clock positions, sending swarms of comets in all directions only during those two relatively brief periods when the planet's aphelion is in or near the plane of the comet disk. If the orbit of Planet X completes one precession cycle in 56 million years, it will cause two periods of comet showers for the inner solar system—one every 28 million years.

But what would be the size of the orbit of a planet with an orbital precession of 56 million years? To their surprise, say Whitmire and Matese, the average distance of this planet from the Sun turned out to be only about 80 astronomical units (about twice the distance of Pluto). Its 700-year orbit would be substantially inclined (perhaps 45 degrees) and elliptical (slightly more than Pluto's). It would range far enough to chop through the comet disk when precession placed it in the proper position, but it was also close enough to the Sun to be detectable. The orbit of this Planet X would be stable.

It was at this point, say Matese and Whitmire, they noticed that their hypothetical planet was practically identical to the one proposed by Harrington and Van Flandern to explain a completely different set of problems. Harrington and Van Flandern's tenth planet was intended to solve

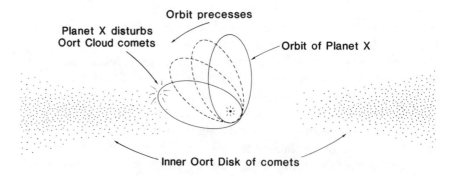

According to Daniel P. Whitmire and John J. Matese, Planet X has an elliptical orbit which is highly inclined to the plane of the solar system. The precession of that orbit causes it to pass through the disk of the inner Oort Cloud of comets, disturbing their orbits and causing comet showers for the inner solar system. Comet impacts on Earth from these showers, they believe, are responsible for periodic mass extinctions of life on our planet.

the irregularities in the positions of Uranus and Neptune, the strange orbits of Neptune's satellites, and the peculiar orbit of Pluto. Whitmire and Matese felt they had found a tenth planet in their data. Harrington and Van Flandern felt they had found a tenth planet in their very different data. Yet the two planets were essentially the same. Matese and Whitmire found in this convergence a great boost to their confidence.

They agree with Harrington and Van Flandern that there is a trans-Plutonian planet of substantial size to be found. No object in the solar system has caused more trouble. Imagine a Most Wanted poster for this fugitive planet.

WANTED on cosmic charges:

- Disturbing the motion of Uranus and Neptune.
- Smuggling short-period comets (like Halley's) into the inner solar system.
- Suspected of trespassing at Neptune, driving Triton and Nereid berserk, and kidnapping Pluto.
- Repeated assaults on Earth with deadly comets, causing periodic mass extinctions of life.

DESCRIPTION of fugitive: One to five Earth masses; eccentric, with odd inclination; likes to leave subtle clues to tantalize astronomers; lives in trans-Plutonia, constantly on move, no known address, might repeat movements every 700 years; knows how to hide.

NOTE: Substantial reward for information leading to his arrest.

Finding one planet, in the opinion of Whitmire and Matese, could solve many crimes.[9]

An analysis by Eugene M. Shoemaker and Ruth F. Wolfe, however, argues that even with a mass of five Earths, Planet X could not cause comet showers, a possible cause of mass extinctions on Earth. Pointing to other studies, they concluded that the purported inconsistencies in the motions of Uranus and Neptune do not indicate a tenth planet, that a planet of equal or greater mass than Earth would be so bright that it could hardly have escaped previous discovery, and that a planet larger than Earth would be unlikely to have formed beyond Pluto.[10] Matese and Whitmire counter that their own analyses show that a Planet X with five Earth masses could cause comet showers, that a planet with low reflectivity or one located deep in the southern sky could have escaped detection, and that the study that indicated an Earth-size planet could not form beyond Pluto also cannot account for the existence of Uranus and Neptune.

Watching for a
Spacecraft
to Quiver

John Anderson
Courtesy of John Anderson

In 1987 a new proponent of a tenth planet emerged on the scene. John Anderson at the Jet Propulsion Laboratory was working with NASA's *Pioneer 10* and *11* spacecraft, which had preceded the *Voyagers*. *Pioneer 10* flew by Jupiter. *Pioneer 11* flew by Jupiter and converted its close passage into a gravity assist to Saturn. Both craft are now headed out of the solar system in nearly opposite directions, and *Pioneer 10* is already beyond the orbits of the planets.

Anderson tracked the departing probes by their radio signals for five years to see if either was deflected from its course by gravity waves moving across the cosmos or by the gravity of a trans-Plutonian planet. Neither craft exhibited any sign of unusual perturbation.

Anderson therefore concluded that a tenth planet most likely does exist! A positive conclusion from what seemed like negative evidence.

Two years earlier, the interpretation of this data by Anderson and E. Myles Standish, Jr., had been negative: "A three-year analysis of radio tracking data from the Pioneer 10 spacecraft . . . fails to reveal the presence of Planet X."[11]

But now Anderson felt that the slight unpredictabilities in the motion of the outer planets that U.S. Naval Observatory astronomers had called attention to could best be explained by the existence of a tenth planet. Therefore, the lack of perturbations in the spacecraft provided information about the unseen planet's path.

The Jet Propulsion Laboratory had calculated highly accurate orbits for Uranus and Neptune by including only observations from 1910 to the present and by excluding earlier observations. Anderson had confidence in almost all the observational data—new and old. He therefore concluded that an unseen outer planet had been disturbing Uranus and Neptune before and during the nineteenth century but that the planet was now too far away for its gravitational effects to be noticeable on the planets or even on the two tiny and distant *Pioneer* craft.

Thus Planet 10, he surmised, must have a highly elliptical orbit that carries it far enough away to be undetectable now but periodically brings it close enough to leave its disturbing signature on the paths of the outer planets. He put the orbital period at 700 to 1,000 years. The suspect planet also had to have a highly inclined orbit so as to have produced no detectable deflection of either of the *Pioneers*. Finally, to create the planet perturbations reported in the Naval Observatory data, he concluded that the planet must have a mass of about five Earths. This Planet X is now far away, nearing aphelion, where its gravitational effects are unnoticeable. Its perturbations on the outer planets won't be detected again until about the year 2600.

For the time being, this distant planet, if it exists, has left us few, if any, clues as to where in the sky to search for it. Anderson hopes that, as the two *Voyager* craft sail out beyond the planets, they may help to pin down the location of Planet X.

The most recent of the proposals for a tenth planet is also the most traditional in methodology. Conley Powell is an aerospace engineer for Teledyne-Brown Engineering in Huntsville, Alabama, who specializes in astrodynamics. He calculates orbits. So slight discrepancies in the motions of the outer planets that U.S. Naval Observatory astronomers could not explain as observational errors were fascinating to him. He discarded the Neptune residuals from consideration even though Harrington and Van Flandern thought them significant because he felt that, with less than one full revolution completed since it was discovered, Neptune's precise orbit was still too uncertain. The orbit of Uranus, however, was much better known, and he began an analysis of the discrepancies between the predicted and actual positions of Uranus in the tradition of Adams, Le Verrier, and Lowell, with the hope that the position of a tenth planet could be determined mathematically.

As the Jet Propulsion Laboratory mathematicians had shown, the observations of Uranus after 1910 provide much more orbital accuracy than early ones. Powell therefore assigned less weight to early visual and even early photographic observations and more weight to twentieth-century Uranus position records, based on probable errors in the data as computed according to statistical theory. It was not, he felt, a question of the disturbing planet getting farther away during the past 80 years. The data improvement about 1910 was too sudden to be attributed to the growing distance of a tenth planet. It was due instead, he thought, to new techniques used to pinpoint planet positions.

Powell's calculations showed Planet X with 2.9 Earth masses at a distance from the Sun of 60.8 astronomical units, giving a period of revolution of 494 years.[12] He was intrigued that this number was approximately twice the period of Pluto and three times the period of Neptune— suggesting that the planet he thought he saw in the data had an orbit stabilized by mutual gravitational resonance with its nearest neighbors despite their vast separation.

His calculations for a tenth planet also provided an orbit inclined by only 8.3 degrees (less than Pluto's) and only slightly eccentric (but not firm enough to warrant a number). The orbital resonance of Planet 10 and the unspectacular nature of its derived orbit gave him confidence that this unseen planet was real.

This solution to his calculations called for the trans-Plutonian planet

Reanalyzing the Motion of Uranus

Conley Powell
Courtesy of Conley Powell

to be in Gemini, ironically the same part of the sky where Uranus and Pluto were found. It would be brighter than Pluto when it was discovered.

At the request of Conley Powell, the Lowell Observatory agreed to search for the hypothesized planet near the spot that Powell's calculations indicated on the survey photographs made by Clyde Tombaugh. Had Tombaugh missed the tenth planet? Beginning in August 1987, under the supervision of Edward Bowell, Norman Thomas used a blink comparator to search approximately 1,000 square degrees of sky. Nothing was found.

Undaunted, Powell reexamined his data. He was dissatisfied with his inability to derive more than a vague value for the eccentricity of the orbit of his suspected tenth planet. So he let his computer "have its head" in a sequence of approximations that produced a still better fit of the residuals. What emerged was a significantly different prediction.

Planet 10 was now smaller, closer, and brighter. His new solution predicted a planet with 87 percent the mass of Earth. With a mean distance of 39.8 astronomical units, a period of 251 years, and an eccentricity of 0.26, Powell's planet would follow almost exactly the path of Pluto. Like Pluto, it would even spend part of its circuit inside the orbit of Neptune. Powell is fascinated by the similarities between his new prediction and the final predictions for a ninth planet by Lowell and Pickering.

This tenth planet, Powell calculates, should be located at present in the eastern part of the constellation Leo, the Lion, with a comparatively bright magnitude of 12.2. "It's hard to see how Tombaugh could have missed it," says Powell. Although the unseen planet would have been well toward aphelion during Tombaugh's search, it would have been close to the ecliptic and brighter than magnitude 13. Says Powell, "Either Tombaugh was mistaken in thinking that he could not have missed a planet brighter than 16th magnitude or my calculations are mistaken."

Powell thinks that a search for the planet predicted by his revised computation is premature. He first wants to refine his calculations using new data he has just received from Bob Harrington—the post-1910 residuals of Uranus calculated by the Jet Propulsion Laboratory for navigational use in the *Voyager 2* encounter with Uranus.

If his planet is found, he would like to name it Persephone. But if his calculations lead to no planet discovery, would that suggest that the Uranus residuals are erroneous? No, says Powell, it is hard to believe that these discrepancies are not significant. More likely, he feels, his gravitational model of the outer solar system is wrong, perhaps because there is an eleventh planet out there whose gravitational effects are confusing matters.

There is yet another serious scientific search in progress for a trans-Plutonian planet, but it is being conducted without the belief or theoretical calculations that such a planet really exists. This investigation is being carried out by Thomas J. Chester and Michael Melnyk of the Jet Propulsion Laboratory. It is perhaps the least orthodox of all.

On January 25, 1983, NASA launched the Infrared Astronomical Satellite (IRAS), a joint endeavor by the United States, the Netherlands, and Great Britain. Scientists realized that this 22-inch (57-centimeter) telescope with infrared sensors would have a reasonable chance of detecting a tenth planet if it exists. A distant planet would receive and reflect very little sunlight and therefore be hard to find with optical telescopes and photography. But the meager energy it absorbed from the Sun and converted to heat could make this planet stand out clearly in infrared wavelengths. Many research projects were conducted on IRAS before it ran out of liquid helium to cool its special infrared detectors in November 1983, as expected.[13] Regardless of whether they were examining galaxies, nebulae, or stars, many of the astronomers checked the IRAS data for a possible tenth planet. As Chester explained, "Everyone wanted to be rich and famous."

It wasn't easy. IRAS recorded some 600,000 objects. Finding a tenth planet among them is like finding a poppy seed in a bin of sesame seeds. Almost. There were some sensible ways to improve the odds. As they proceeded with their principal investigations, astronomers kept their eyes open for objects in the expected low-temperature range. They looked for objects that moved. No tenth planet was found.

Looking for Heat

Thomas J. Chester and Michael Melnyk (right)
Gaylin Laughlin

Plate 16. Saturn and its exquisite ring system as seen by *Voyager 2*. The two bluish dots below Saturn are its moons Rhea and Dione.
NASA/Jet Propulsion Laboratory

Plate 17. Even more detail in the ring system of Saturn is visible in this *Voyager 2* view after closest encounter. Saturn can never be seen from this angle on Earth.
NASA/Jet Propulsion Laboratory

Plate 18. The *Voyager* spacecraft revealed that Saturn's ring system is composed of thousands of rings, as this false color picture clearly shows.
NASA/Jet Propulsion Laboratory

Plate 19. The richly banded world of Jupiter, as seen by *Voyager 1*. The Great Red Spot, a giant storm more than twice the size of Earth, and many lesser storms (white and reddish spots) are clearly visible. Also visible are three moons of Jupiter: reddish Io is near the right edge of Jupiter; Europa is the whitish dot near the right edge of the picture; and Callisto is a tiny dot below Jupiter and to the left.
NASA/Jet Propulsion Laboratory

Plates 20 and 21. Io, Jupiter's innermost large moon, as photographed by *Voyager 1*. Active volcanoes on Io, such as the one seen on the horizon (Plate 21), hurl gas and debris more than 100 miles (160 kilometers) above the moon's sulfur-covered surface. Said one scientist: "I've seen better-looking pizzas."
NASA/Jet Propulsion Laboratory

PLATE 16

PLATE 17

PLATE 18

PLATE 19

PLATE 20

PLATE 21

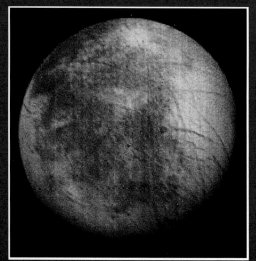

PLATE 22

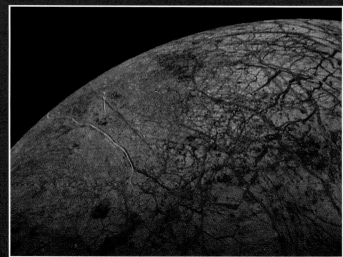

PLATE 23

PLATE 24

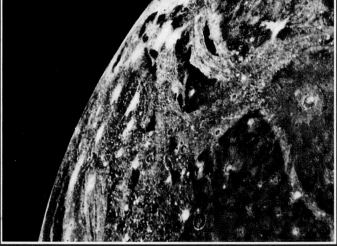

PLATE 25

PLATE 26

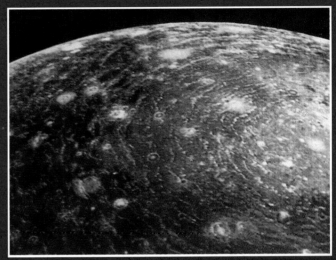

PLATE 27

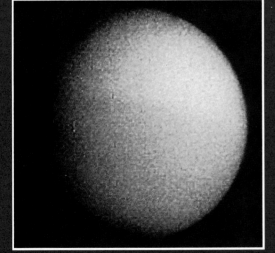

PLATE 28

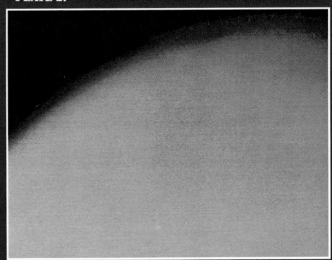

PLATE 29

PLATE 30

Plates 22 and 23. Europa, just a little smaller than Earth's Moon, looks like a fractured billiard ball in these *Voyager* pictures. Beneath the icy surface may lie liquid water. The fractures in Europa's crust are probably caused by the tidal strain created by Jupiter, Io, and Ganymede.
NASA/Jet Propulsion Laboratory

Plates 24 and 25. Ganymede, Jupiter's largest satellite and the largest moon in the solar system. The ridges and grooves seen by the *Voyagers* indicate a world where chunks of crust moved about and collided with one another.
NASA/Jet Propulsion Laboratory

Plates 26 and 27. Callisto, the outermost large moon of Jupiter. *Voyager* pictures showed that it was saturated with craters. Any new craters formed by crashing asteroids or comets will obliterate older craters. This thoroughly cratered landscape suggests that Callisto has changed little since it formed.
NASA/Jet Propulsion Laboratory

Plate 28 and 29. Titan, Saturn's largest satellite and the second-largest moon in the solar system. Only two moons, Titan and Neptune's Triton, have permanent atmospheres. The *Voyagers* found that the principal gas in Titan's atmosphere is nitrogen and that the atmospheric pressure is 60 percent greater than that of Earth.
NASA/Jet Propulsion Laboratory

Plate 30. Around the star Beta Pictoris is a disk of matter that may be icy planetesimals left over from the formation of a planetary system. Several other stars are now known to possess such disks. Although planets, if any, are not detectable, the resemblance of this disk to the hypothesized inner Oort Cloud of our solar system suggests that the formation of planets is common throughout the universe. Bradford A. Smith and Richard Terrile blocked out most of the light from the center star so that the much fainter disk material could be seen.
Jet Propulsion Laboratory

Infrared Astronomical
Satellite (IRAS), a joint
endeavor of NASA, the
Netherlands, and Great
Britain, was launched
by NASA in 1983 and
performed superbly for
nine months until, as
expected, its coolant
was exhausted. With its
22-inch (57-centimeter)
telescope and its ability
to study infrared radia-
tion absorbed by the
Earth's atmosphere,
this instrument pro-
vided views of the
heavens never before
available. Astronomers
are still combing the
observations for
discoveries, including
the possibility of a
tenth planet in our
solar system.
*NASA (photograph and
artwork)*

There was only one method left: Cull the entire IRAS catalog of objects for tenth planet candidates—cool, faint sources—and compare every candidate to Palomar Sky Survey optical photographs taken almost 40 years ago. If the IRAS catalog and the survey photo both showed the object at the same position, then it *wasn't* Planet 10 because the planet would have moved. This systematic search was the approach that Chester and Melnyk pursued. Fortunately, not every star had to be compared to an optical survey photograph. Existing catalogs listing the positions of infrared and optical objects could be used to disqualify 80 percent of the stars seen by IRAS. If the IRAS object appeared in the catalog, it hadn't moved and was eliminated.[14]

Chester and Melnyk are gradually compiling a list of a hundred or so most suspicious objects to be checked out by the 200-inch Hale Telescope on Mount Palomar. A few candidates have already been examined, but no Planet 10 has been found.

These two IRAS gleaners have simplified their search one step further. They have concentrated on the portion of the heavens where there are the fewest stars. In this way they can cover the most sky with the least effort. They have avoided the plane of our Milky Way Galaxy, where the stars are most concentrated, and have started with the galactic poles. Since 1986 they have been working their way from the north galactic pole to a galactic latitude of 50 degrees. So far, they have covered one-tenth of the entire sky. They think they can cover down to nearly 30 degrees before star density is too great. When they reach that point, they will have examined 120,000 IRAS objects. So far they have completed 20 percent of their task. Chester explains: It's like a drunk looking for his lost keys at night around the base of a lamp post. A passerby stops to help. "Is this where you think you lost them?" he asks. "No," says the drunk. "But the light is better here."

Thus the search by Chester and Melnyk is not based on a planet perturbing Uranus and Neptune (Chester doubts that the residuals are due to a tenth planet). The search is not based on a planet periodically dislodging comets to deluge the Earth (Chester believes the evidence for that hypothesis is extremely weak).

Instead the Chester and Melnyk review of IRAS data for a tenth planet is a by-product of Chester's principal mission—to find objects outside the solar system that have too little mass to shine like a star by nuclear fusion. These objects, more massive than Jupiter but less than 8 percent the mass of our Sun, glow feebly because of the heat generated by the gravitational compression of their interiors and the radioactive decay of heavy elements trapped within them. Even if these so-called brown dwarfs were closer to our solar system than the nearest stars, they may be impossible to see with optical telescopes.[15] But if they exist, they must be visible in the infrared and thus should have been captured in the IRAS data. Chester is trying to see how much undetected matter there is in the form of brown dwarfs in our vicinity.

The amount of unseen matter has important implications for the future of the cosmos. The universe is currently expanding following its origin in the big bang 15 to 20 billion years ago. Will the universe continue to expand forever, or will the expansion gradually slow to a halt and begin a contraction that will lead over tens of billions of years to a universe compressed into a singularity, perhaps followed by a new big bang? The answer lies in whether there is enough gravity to halt the expansion, which means whether there is enough mass in the universe.

Telescopes have detected only about 10 percent of the necessary density to keep the universe from expanding forever. Is there a vast amount of unseen matter, perhaps as black holes, neutrinos, clouds of gas and dust, very faint stars, or substellar objects? If our corner of the universe contains a modest number of brown dwarfs, and if our locale is reasonably typical (it seems to be), then brown dwarfs nearby would indicate that the universe is denser than we have previously detected, which would increase the chances that we live in an oscillating universe. And if IRAS has found very few or no brown dwarfs, that may be important evidence that brown dwarfs make little or no contribution toward the missing mass needed to "close" the universe. Perhaps, instead, we live in an open universe, its expansion to continue forever. Regardless of how many brown dwarfs are found in the IRAS data, as long as Chester interprets the data properly, his findings should have important implications for one of the fundamental questions about the universe.

The IRAS search may also detect previously unknown types of celestial objects, which is Melnyk's major motivation for undertaking this project.

As for a tenth planet, what could IRAS expect to find? Its sensitivity and resolution were not ideal for a tenth-planet search, but it was a powerful tool. For example, it could detect Pluto at its present distance, yet just barely. A Pluto-size planet farther away would be invisible to IRAS. But Planet 10 hypothesizers are not looking for something that small. They are searching for something massive enough to be disturbing Uranus and Neptune or massive enough to perturb great numbers of comets. Here IRAS could be of help. It could detect a Jupiter-size planet at a distance of 5,000 astronomical units—more than a hundred times farther out than Pluto ever ventures. Most concepts of Planet 10 envision a body between the size of Earth and Uranus. Depending on the internal heat such an object might generate, IRAS could feel it at a distance of 60 to 250 astronomical units.

For Chester and Melnyk, the discovery of a tenth planet would be a pleasant and rather unexpected surprise. But then, says Melnyk, IRAS has provided many serendipitous discoveries. If they do find a tenth planet, Melnyk thinks that Cronus might be a good name, since he was the father of time and a distant new planet would certainly take lots of time to make its way around the Sun. Their failure to find one would not rule out a tenth planet but would further constrain the possible size and distance of such an object and the part of the sky where it might be hiding. If Chester and Melnyk find no tenth planet, they feel they will be able to say with great confidence that no tenth planet of significant size and proximity exists in the 50 percent of the sky they have covered. The unsearched sky remaining will be close to and along the Milky Way, where the stars are so numerous that a complete search of the IRAS data performed by human beings would be so lengthy and tedious that it would be impractical.

Expanding the Tombaugh Search

The similarities in conceptions of a tenth planet expressed by Harrington and Van Flandern, by Whitmire and Matese, and by Anderson do not necessarily point to the existence of an undiscovered planet. To the extent that predictions of a tenth planet are based on the irregularities in the motions of the outer planets being real rather than errors in observation, transcription, or calculation, the missing planet can have only a rather modest mass because the purported discrepancies in the motions of Uranus and Neptune are so small. Concepts of a tenth planet are also constrained by the observational work of Clyde Tombaugh and others engaged in subsequent surveys. To have avoided detection, the tenth planet must be distant and faint, most likely in a highly elliptical orbit so that it is at present far from the Sun and in a highly inclined orbit so that it is far from the plane of the solar system where the most intensive surveys have been conducted.

Clyde Tombaugh's careful search before and after the discovery of Pluto is still so greatly respected that few serious proponents of a tenth planet of 16th magnitude or brighter think that it could be undetected in the 70 percent of the sky surveyed by him. And to Tombaugh's survey has been added the work of Charles T. Kowal.

In 1977, Kowal began a new systematic search for undiscovered bodies in the solar system using the Palomar Observatory's 48-inch (122-centimeter) Schmidt Telescope, a photographic survey instrument with 14 times the light-gathering power of the 13-inch (33-centimeter) astrograph that Tombaugh used to find Pluto. In the course of his seven-year search covering 15 degrees above and below the ecliptic, Kowal found 5 comets and 15 asteroids, including Chiron, the most distant asteroid known. He also recovered 4 lost comets and 1 lost asteroid.[16] He did not find a tenth planet.

Charles T. Kowal with the 48-inch Schmidt Telescope at Palomar Observatory that he used in his survey for new objects in the solar system.
Courtesy of Charles T. Kowal

He concluded that there was no planet brighter than 20th magnitude within 3 degrees of the ecliptic. He could have detected a Jupiter-size planet out to a distance of 800 astronomical units, about 20 times Pluto's mean distance from the Sun. He is doubtful that any sizable planet lies within the 30-degree belt of his survey, although he points out that the search took a long time and objects could have eluded him by moving out of a to-be-photographed area into one previously examined. Kowal says he wouldn't be surprised if a new planet with a high orbital inclination were discovered. He notes that when Pluto is at maximum distance south of the ecliptic (17.2 degrees), it would have been outside his field of search.

Kowal himself has made an unpublished attempt at predicting the location of a tenth planet and has come up with numbers very similar to those of Harrington and Van Flandern, Powell, and others—a planet several times the mass of Earth and bright enough so that Tombaugh should have discovered it. "I spent many years of my life blinking photographic plates," says Kowal, "and I know how easy it is to miss something."

Of his calculations, Kowal notes: "I obtained perfectly reasonable predictions of the orbit of the unknown planet by using the residuals of Uranus and the residuals of Neptune. The only problem is, the two predictions do not agree with each other." He hopes to return to this work soon.[17]

After seven years at the telescope and blink microscope, Kowal formed an indelible conclusion: A general search for very faint planets such as he was conducting is not feasible. It takes too much time and exhausts human and material resources. If a tenth planet exists, astronomers must calculate its approximate position and tell observers where to look. Kowal's systematic search for undiscovered bodies in the solar system was abandoned in 1984.

The searches of Tombaugh, Kowal, and others have greatly restricted the regions of the sky where a tenth planet is thought likely to be hiding. But what about the theoretical underpinnings of a tenth planet, especially those taunting discrepancies between the predicted and actual positions of Uranus and Neptune?

Problems with the Data?

Brian G. Marsden is director of the International Astronomical Union's Central Bureau for Astronomical Telegrams, astronomy's clearinghouse for discoveries of comets, asteroids, moons, supernovae, quasars, pulsars, and planets. He is an expert on orbital computations. He thinks that problems in fitting Uranus and Neptune to orbits may be due to the way the data were taken and that too much reliance has been placed on the accuracy of older prediscovery observations, such as Lalande's two glimpses of Neptune in 1795 and Galileo's plotting of Neptune in 1613. Remember, he cautions, Neptune has not yet completed one revolution around the Sun since it was discovered and Uranus is only halfway through its third. It is too early to rule out errors of observation or data transcription as an explanation for the disparities. The orbits are not yet known with enough precision to permit the conclusion that the gravity of an undiscovered planet is disturbing them.

The extraordinary flight of *Voyager 2* beyond Jupiter and Saturn converted the problems in the motion of Uranus and Neptune from a matter of theoretical interest to a matter of practical urgency. To fly a spacecraft within a few miles of a distant planet's cloud tops and moons and use that planet for a gravitational assist to a yet more distant planet, it is necessary to know precisely how both those planets are moving and exactly where they will be.

E. Myles Standish, Jr., and his colleagues at the Jet Propulsion Laboratory faced that challenging problem. The precision of their predictions would have extraordinary consequences. They recalculated the orbits of Uranus and Neptune. The ephemeris they created for Uranus very accurately predicted the position of the planet when *Voyager 2* arrived. They hope their ephemeris for Neptune will allow *Voyager 2* to be equally successful when it reaches its final target. The celestial mechanicians achieved their results, however, only by ignoring observations of the positions of Uranus and Neptune prior to 1910.

Because Uranus was right where the new ephemeris predicted it would be even though it was based on less than one revolution of Uranus, Standish doesn't see any need to explain discrepancies in the motions of the outer planets by a tenth planet. He strongly suspects that over the centuries there have been a few undetected errors in the observations and

GALILEO'S OBSERVATION OF NEPTUNE AND ITS IMPLICATIONS
by Dr. E. Myles Standish, Jr., Jet Propulsion Laboratory

Nearly two and a half centuries before the discovery of Neptune in 1846, the astronomer Galileo unknowingly observed that planet through his telescope, noted its motion in the sky, and drew a diagram of its location. This startling fact was uncovered in 1980 by Charles T. Kowal and Stillman Drake while searching through Galileo's notebooks.

Galileo drew the object as a background star for reference. There is no question, however, that it was Neptune. Yet the diagram he drew is ambiguous: Different interpretations have profoundly opposite implications.

In the years after his discovery of the four large satellites of Jupiter, Galileo would routinely observe these moons, measuring their positions and drawing the configurations in his notebook. Occasionally, he would also show the relative location of a background "fixed star," indicating the approximate direction and distance to the star from Jupiter with a dashed line.

For his entry on the night of December 28, 1612, there is no star near where the dashed line is pointing. Instead, the line points unmistakably toward the planet Neptune, the first known sighting of this distant planet. It is likely to remain so. Neptune can be seen only with a telescope, and astronomical telescopes had been in use only three years.

Galileo's diagram of Jupiter's moons on January 28, 1613. He unknowingly plotted Neptune as a background star

Galileo saw Neptune again on January 28, 1613 (shown here in the diagram he made). Jupiter and three satellites are connected with a solid line, and their distances (expressed in units of Jovian semi-diameters) are labeled numerically. A dashed line runs from Jupiter to the lower left edge of the page even more in their handling. The orbit of Uranus has been calculated from many different observations using many different reference systems. In converting these positions to a common reference frame, errors could easily be made. Some checks show conversion errors of 2 arc seconds or more, about four times the amount of the discrepancies in the positions of Uranus that were cited as an indication that it is being disturbed by a distant planet.

Standish strongly advises against the use of prediscovery sightings of Uranus and Neptune because when those positions are discarded and calculations like those of Adams and Le Verrier are made on the remaining residuals, the need for Planet X decreases dramatically. The discrepancies that remain are small enough to be explained by observational error.[18]

where there is a star shown as an asterisk labeled *a*. The distance of *a* from Jupiter is labeled as 29. The dashed line is continued at the lower right, coming to two asterisks, labeled *a* and *b*, but with no numerical measurement given for their separation. On the page Galileo has written, "Beyond fixed star *a* another followed in the same line, as [does] *b*, which also was observed on the preceding night; but they [then] seemed farther apart."

The Smithsonian Astrophysical Observatory catalog identifies *a* as star #119234. But there is no star anywhere near the location of *b*; Neptune was there instead. Galileo's diagram for January 28 also contains an unusual first-time entry: a solid line below Jupiter with two hash marks and the statement, "This is an exact scale of 24 semidiameters."

Present-day computations show that in this diagram, the satellite measurements are accurate to 0.1 semidiameters or better(!), after accounting for a consistent 9 percent scale factor present in all of Galileo's diagrams. But modern computations show that the distance to star *a*, labeled as 29, should have been labeled 27.4, an error of 1.6, nearly 20 times below Galileo's normal accuracy. On the other hand, if we use Galileo's own scale of 24 to measure Jupiter's separation from star *a*, we get 26.5, an error of 2.5 from the 29 Galileo noted. Clearly, Galileo did not measure this dashed-line distance carefully, nor did he draw it to scale.

The Question: Was the distance between SAO#119234 and Neptune drawn to scale? If so, the distance would have been about 3.6, equivalent to 3.9 when corrected for the scale factor of 9 percent. Modern calculations for the position of Neptune on January 28, 1613, put it at a distance of 6.9—a very different number.

The Arguments: The first-time presence of the 24 semidiameter scale and the mention that one of the background fixed stars seemed to have moved overnight indicate that Galileo wanted to follow these objects further; he may have drawn the separation of the asterisks to scale. On the other hand, Galileo consistently labeled his measurements with a numerical value, using a solid line when drawing to scale and a dashed line when indicating mere direction. He used a dashed line here. Unfortunately, Galileo did not again observe the two objects, presumably because of clouds over the next few days.

The Implications: If Galileo's diagram is indeed drawn to scale, as Kowal and Drake believe, then the presence of a tenth planet in the solar system is almost a certainty. The modern ephemerides of Neptune cannot contain an error as large as 3 Jovian semidiameters (more than one arc minute). However, if Galileo did not bother with scale and drew the asterisks only to show the alignment with Jupiter, then his diagram demonstrates the accuracy of modern planet position tables—which are based on the gravity of only nine planets.

Standish has begun an analysis of the older residuals and has noted irregularities in the observations of all the planets. "Did Planet X visit each one on a grand tour?" he asks pointedly. He has also detected differences in the irregularities calculated by the U.S. Naval Observatory, the Royal Observatory Greenwich, and the Paris Observatory. He hopes very soon to try to reconcile the pre-1910 observations with the modern orbits by going back to the raw data at the Royal Greenwich Observatory, the Paris Observatory, and the U.S. Naval Observatory and carefully converting the positions to a common reference system.

There is no way to absolutely rule out the existence of a tenth planet, he notes, but there is no need for one to exist based on the motion of the planets. And new planet-position measuring techniques offer hope of great refinements in our knowledge of precise planetary orbits.

For the past two years, the Very Large Array of radio telescopes in Socorro, New Mexico, has been observing the planets at radio wavelengths. The twenty-seven 82-foot (25-meter) dish antennas move along railroad tracks to separations as great as 23 miles (37 kilometers), allowing the telescope array to resolve the direction of radio signals to an ac-

---

PITFALLS IN PREDICTING A TENTH PLANET
by Dr. E. Myles Standish, Jr., Jet Propulsion Laboratory

The residuals of Uranus and Neptune do not demand the existence of a tenth planet, for there are other explanations that are at least as plausible. It is easy to ignore the fact that the residuals themselves need reexamination.

Do the residuals represent true deviations in the motion of the planets? Or can the residuals be more simply explained as errors and inaccuracies in the procedures of working with the observational measurements?

There is a succession of steps involved in determining the residuals for a planet:

1. Begin with the raw measurements. Classical planetary position observations are made by measuring the altitudes of both the planets and the stars and timing their transits across the meridian (the north-south line in the sky). This step can never be repeated for a specific observation. One cannot go back and reobserve the position of Mars in 1850.

2. Derive the observed positions, O. Using the raw measurements and the most accurate catalog of star positions available, derive the observed positions of the planets. These *derived* "observables" have been published and, more recently, have been put into computer-readable form. Many of the *raw measurements* have also been published, while others still exist only in the original observing notebooks. These raw measurements have not, however, been put into computer-readable form—a task that would be colossal.

3. Calculate the computed positions, C. Use an existing orbit (ephemeris) to predict positions of the planet for comparison with the observed positions.

4. Form the residuals, $O-C$. The residuals are the differences between the observed orbital positions, O, and the computed orbital positions, C. In the ideal world, with

curacy of 0.02 seconds of arc—about 20 times better than standard positional measurements by telescopic photography in the optical wavelengths.

Pushing beyond that accuracy are teams of astronomers at the California Institute of Technology's Jet Propulsion Laboratory and NASA's Goddard Space Flight Center. They are using very widely separated radio telescopes to determine the positions of quasars to an accuracy of 0.001 arc second. Quasars are the most distant objects in the universe and therefore are the closest thing the cosmos offers to a fixed background of objects that do not appear to move up, down, or sideways. By relating spacecraft and planets to the reference background of quasars, navigation of space probes and the predicted positions of planets should be improved by a further factor of 20.

Thinking back over his discovery of Pluto and his 14 years of planet searching, Clyde Tombaugh offered Ten Special Commandments for a Would-Be Planet Hunter. The final commandment decrees: "Thou shalt not engage in any dissipation, that thy years may be many, for thou shalt need them to finish the job."[19]

---

perfect observations and with perfect ephemerides, the residuals would be zero.

5. Adjust the computed orbit of the planet, C, to fit the observed positions, O, as nearly as possible, so as to minimize the residuals. (This adjustment is usually accomplished using the mathematical method of least squares, developed by Gauss.)

6. Examine the final residuals. What keeps them from being zero? Perhaps the catalog of star positions still has distortions. Perhaps the computed orbit can be further improved. When no more improvements are possible, one goes on to step 7.

7. Other explanations of the residuals. What can cause nonzero residuals to a best-fit orbital calculation? A defect in the law of gravity? Planet X?

Star catalogs produced since 1910 have been shown to contain significant distortions. Certainly earlier and cruder catalogs must also contain distortions of similar, if not greater, magnitude. However, no one has ever gone back to step 2, properly reprocessing the raw measurements of the planets using a modern star catalog.

For the post-1910 data, a lesser alternative has been used: updating the published positions using general differences found by comparing the original catalog with a modern one. For data previous to 1910, not even this lesser alternative has been applied. The original catalog distortions remain embedded in the residuals.

At best, a few tenth-planet seekers start at step 5, cranking out new orbits in an attempt to fit the observations. Far worse, most begin with step 7: "May I get a copy of your residuals?"

The complete job must start back at step 2—taking the original measurements of the planets and recalculating their positions using modern star catalogs. Such an analysis would bring no funding or headlines. It would be time-consuming and tedious. But until the basic observations (from which come the residuals) are processed properly, there is no necessity to invoke Planet X.

CHAPTER 14

Cosmic Archives:

What the Outer Solar System Can Tell Us

"The worlds of the outer solar system, with their odd characteristics, do not require some grand conspiracy. What is required is an appreciation of the countless objects that once plied those dark reaches."
Astronomer William B. McKinnon
(1988)

On Earth, a paleontologist digs down through recent fossils to ever-older specimens. The deeper he digs, the further back in time he goes. For the planetary scientist who wants to probe back in time to the beginning of the solar system to learn of the conditions and the chemicals present when the Sun and planets formed, the route is not down but outward.

Because of its weather and tectonic activity and because of its position so close to the Sun, the Earth has erased or lost its earliest history. The Moon, smaller in size, retains material less changed from an earlier era but cannot carry us back to the beginning. The meteorites that crash on Earth, fragments of minor planets from the asteroid belt, are the most pristine representatives of the bodies from which the inner planets took form. Yet they have been scoured by sunlight and solar particles and have lost their lighter elements. They stop short of revealing the solar system's beginnings.

The gas giants Jupiter and Saturn, however, do preserve part of that information. Their compositions, so similar to the Sun, tell us something of the cloud of gas and dust from which our solar system coalesced. But their masses have engendered such internal heat that they have changed considerably from that primordial cloud.

The farther out we go, the more primeval conditions become—the cold and dark before our Sun began to shine. Here we reach the horizon of ice and bodies too small to have induced great change within themselves. Here, if anywhere, lie the clearest, earliest clues of our origins—and perhaps the origins of other planetary systems that we cannot see directly but which we suspect surround the stars in great profusion.

In the comets, we may have planetesimals, the small accretions of ice and flecks of rock from which all the larger bodies throughout the solar system—the major and minor planets and the moons—began to form. In Pluto and Charon and perhaps in Triton and Nereid, we may have the step that followed planetesimals in planet construction—bodies grown more than large enough by gentle adhering collisions of planetesimals to enhance their growth by gravitational attraction. Then, in Uranus and Neptune, we find monuments constructed from trillions of cometlike planetesimals, so large that hydrogen, helium, and other gases from the solar nebula collapsed onto them by gravity to build giant planets with vast atmospheres dominated by light gases. For those who wish to study the early days of the solar system, here is the realm to explore—the most nearly pristine remnants of the stages in the birth of the planets.

Molding Suns and Planets

Nearly 5 billion years ago, an interstellar cloud of gas and dust crossed a threshold. For millions of years, it had gradually been accumulating material that was drifting in the vast reaches between the stars until it was so massive that it began to condense—to pull itself together by gravity. This large nebula fragmented into numerous smaller ones. From each large fragment of this nebula would come a single star or, more often, pairs or families of stars and the small nonluminous bodies we call planets. Several hundred stars would form from this one cloud of gas and dust. Gravity would determine not only which of those fragments would spawn stars and how many but also what fragments had too little mass or the wrong spin to make a family of stars and would have to settle for just one star, surrounded instead only by small dark planets.

In a very modest region of this large interstellar cloud of gas and dust, the density was high enough so that it too began to condense. From this portion of the great cloud would emerge our solar system. Within the solar nebula, each atom and molecule was pulling by gravity upon all the others so that the particles pulled themselves closer together. The solar nebula, at first amorphous, began to shrink and become still denser, shrinking more. All the atoms and molecules of the cloud were in motion in all directions so that the sum total of motion almost balanced out but not quite. The collapsing cloud was just barely turning. But as gravity pulled the cloud in upon itself, it spun more rapidly, so that angular momentum was conserved, like a figure skater spinning faster as she pulls in her arms. The solar nebula not only contracted but also flattened into a disk as the centrifugal force of the rotation kept the particles traveling in the plane of the spin from falling inward.

The more it contracted, the greater the density of the cloud fragment became and the more particles within it collided, creating heat. The outward pressure of the heat slowed the infall of material due to gravity. The cloud glowed red as the temperature within it soared to 3,200°F (1,800°C) or more. The temperature was high enough to vaporize virtually all of the solid material scattered throughout the inner regions of this gaseous cloud.

Contraction ceased in the cloud's outer regions. But at the center, where density was greatest, contraction continued until the temperature of the collapsing sphere reached 18 million degrees Fahrenheit (10 million degrees Celsius) and a nuclear reaction began at the core, changing hydrogen to helium. A star had formed. That star was our Sun. Around it lay a hot disk of matter, the solar nebula.

As the nuclear reactions of the Sun throbbed into life, great gusts of subatomic particles poured forth from our star and swept through the solar nebula, carrying the lighter gases outward.

The nuclear reaction at the core of the Sun balanced the inward pressure of gravity, and the Sun stopped contracting. Its surface was now hotter than the contracting protoSun had been, but the surface area was now far less, so the total energy radiated declined.

As temperatures within the nebula fell, different elements within the cloud condensed to solids and formed compounds with other elements, especially with abundant and reactive oxygen. First the metals condensed, forming solid microscopic grains to which other molecules and grains could adhere. Then silicon condensed in a wide variety of silicate compounds—stony minerals the size of motes of dust. This produced more nuclei around which larger bodies could accrete. Farther from the Sun, where the temperatures were freezing or well below, most of the remaining gases, except hydrogen and the inert elements, condensed to form ices—especially minute snowflakes of water. At great distances, however, the temperature had never risen above freezing, and ice particles there were traces of the nebula from which the solar system was forming.

Arrayed around the Sun, near and far, these flecks of dust and ice were countless tiny solar satellites, each with its own orbit. Collisions were incessant, but since these particles were almost all traveling in nearly parallel orbits in the same direction around the Sun, the collisions were gentle enough for particles to stick together, growing larger. These were planetesimals, the building blocks of planets and moons.

With the ice and dust planetesimals at small size, growth came primarily by collision and adhesion. But with increasing size, gravity began to play a greater role. As they grew still larger, gravity began to predominate. It became a race to gather up mass as fast as possible. The more mass, the more gravity, the bigger the protoplanets grew. Half a billion miles (three-quarters billion kilometers) from the Sun, where temperatures were cool enough for the solid particles to include ice as well as rock, a massive protoplanet took form—and a second at twice that distance. These overgrown planetesimals grew so massive that gases from the solar nebula collapsed upon them by gravity to make them giant planets. These gas giants fought gravitationally with the newborn Sun and one another for control of all the matter lying between them until little was left, and what remained was a satellite of one or the other of the contestants.

The Cosmic Snowball Fight

Beyond these two bodies that would ultimately be Jupiter and Saturn were two more giant planet aspirants. But protoJupiter and protoSaturn had combined their gravities to create a no-planet zone for a great distance beyond. Only at twice the distance of Saturn could one stable body emerge and grow, and then not another until more than that distance again. Even with so much territory to rule and gather tribute from, protoUranus and protoNeptune grew slowly compared with Jupiter and Saturn.

The density of the forming solar system was greatest toward the center and much lower at that distance. And orbital speeds were slow, retarding accretion still further.

The Sun and surrounding stars were partially shrouded by a haze of dusty gas from which all the bodies in the young system were still growing. But now the larger bodies recklessly, helplessly, began to impose a limit of growth upon themselves. They were growing by gravity, absorbing gas and dust along their paths and swallowing up icy planetesimals and even planet-size bodies. Most of the planet population in our original solar system was sacrificed toward the goal of building larger bodies. But not all or even most of the objects in their paths were absorbed. Each planetesimal had its own orbit, and a crash into a large body as they passed near one another was a long shot. The encounters were far more likely to throw the planetesimals onto new orbits, some aimed

AMBASSADOR TO JUPITER: PROJECT GALILEO

by William J. O'Neil, Project *Galileo* Science and Mission
Design Manager, Jet Propulsion Laboratory

In October 1989, the *Galileo* spacecraft is scheduled to be launched on the grandest planetary
mission of the 1990s. Traveling an unusual flight
path, *Galileo* will reach Jupiter in 1995 to become
the first manmade permanent resident of the
Jupiter system.

The *Galileo* spacecraft consists of a probe to sample Jupiter's atmosphere and an orbiter to observe
the planet, its magnetosphere, and its satellites over
a two-year period. This investigation of Jupiter will
be far more comprehensive than was possible with
the *Voyager* and *Pioneer* flyby missions.

When the *Galileo* mission was funded in 1977,
it was planned for launch in January 1982. The
Space Shuttle was to carry it into Earth orbit, where
a special "planetary" Interim Upper Stage (IUS)
would boost *Galileo* onto a direct trajectory to
Jupiter, with arrival in the summer of 1985. Since
inception, however, *Galileo* has been repro-
grammed five times, shifting back and forth between the Centaur liquid hydrogen/oxygen upper
stage and lower-power versions of the IUS. The
Challenger accident resulted in a reduction of Shut-

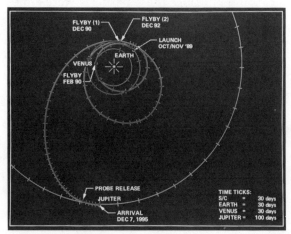

The *Galileo* mission flight path
NASA/Jet Propulsion Laboratory

tle cargo weights and the final cancellation of the
Centaur as an upper stage for the Shuttle (considered an extra risk). Consequently, only the existing two-stage IUS is available to lift *Galileo* into
interplanetary space.

But the two-stage IUS has only enough energy

The *Galileo* probe
descends into Jupiter's
atmosphere while the
orbiter passes above to
relay its findings to
Earth in this artist's
conception.
*NASA/Jet Propulsion
Laboratory*

to send Galileo to Mars or Venus. Because of planetary alignments, a gravity assist from Mars cannot be used to reach Jupiter, but, happily, Venus and Earth can team up to do the job. The *Galileo* trajectory designers discovered that by launching the spacecraft toward Venus in October 1989, it could receive gravity assists from Venus in February 1990 and from Earth in December 1990 and again in December 1992 to achieve a trajectory that will delivery *Galileo* to Jupiter on December 7, 1995.

Galileo, the most sophisticated planetary spacecraft ever built, is the first planet-bound vehicle to use a dual-spin design. The larger portion of the craft spins at 3 rpm to provide attitude stability and a constant scanning motion for the fields and particles instruments. An electric motor spins the lower third of the craft at the same rate but in the opposite direction, which cancels out the rotation of this portion of *Galileo*, thereby making it a stable nonspinning platform for telescopic instruments. Thus *Galileo* has the advantages of both a spinning spacecraft (like the *Pioneers*) and a 3-axis stabilized spacecraft (like the *Voyagers*).

Electrical power for *Galileo* is provided by small nuclear generators. The Federal Republic of Germany, a partner in the project, has provided the spacecraft propulsion system for course and attitude changes. The orbiter weighs 2,800 pounds (1,270 kilograms); the probe 750 pounds (340 kilograms). With a little more than a ton of propellants on board, the launch mass of *Galileo* is almost 3 tons (2,700 kilograms).

Galileo's probe carries six instruments to identify chemicals and their abundances and the energy of particles in the atmosphere of Jupiter. The orbiter has five instruments to measure dust, magnetic fields, and particle energies. It also carries four telescopic instruments: infrared and ultraviolet spectrometers, a photopolarimeter, and an imaging system using a CCD, a special silicon chip with 100 times the light sensitivity of the *Voyagers'* vidicon. In addition, when the *Galileo* orbiter moves behind Jupiter, its radio signal directed toward Earth will pass through Jupiter's atmosphere to reveal its structure and composition.

As originally planned, *Galileo* will collect fields and particles data during its cruise through the interplanetary medium en route to Jupiter. But the now mandatory Venus-Earth-Earth gravity-assist

The *Galileo* spacecraft
NASA/Jet Propulsion Laboratory

(VEEGA) trajectory adds unique science opportunities. During the flyby of Venus and both transits of the Earth-Moon system, *Galileo* will observe these bodies from largely unprecedented angles. The highlight of the interplanetary voyage phase, however, will be the first flyby of an asteroid. Midway between its two Earth gravity assists, on October 29, 1991, *Galileo* will observe the asteroid Gaspra (10 miles [16 kilometers] in diameter) at close range. And, for an encore, *Galileo* will scrutinize the asteroid Ida (20 miles [32 kilometers] in diameter) on August 28, 1993, while flying the final direct leg to Jupiter.

In July 1995, five months before reaching Jupiter, *Galileo* will release its probe to fly untended toward an exacting Jupiter entry at 112,000 miles per hour (180,000 kilometers per hour)—the first manmade object to penetrate the atmosphere of Jupiter. A few days after probe release, the orbiter will be maneuvered so that it will be passing over the probe as it descends into the Jovian atmosphere and can relay the probe's scientific measurements back to Earth. This relay link will be maintained for 75 minutes as the probe sinks to a depth where the pressure of Jupiter's atmosphere is 20 to 25 times greater than air pressure on Earth. At about that level, the probe will be destroyed by Jupiter's environment.

Several hours before the probe's final plunge, the *Galileo* orbiter will make its first close observations of two of Jupiter's four largest moons (often called the Galilean satellites in honor of the great scientist who discovered them in 1610). Europa will be observed at a range of 22,000 miles (35,000 kilometers). Next, the orbiter will fly by Io at an altitude of only 600 miles (1,000 kilometers). Four hours later, the probe data relay will begin. Immediately after the relay, the orbiter will execute its Jupiter orbital insertion maneuver, firing its main engine for 45 minutes to achieve an eight-month initial orbit of Jupiter.

The craft will then begin to utilize gravity assists from the Galilean satellites to bootstrap its way from moon to moon while making an intensive study of Jupiter's turbulent atmosphere and awesome magnetosphere at the same time. On each orbit, the orbiter will be precisely accelerated to the aim point for the upcoming satellite in the sequence. Ten orbits of Jupiter will be completed during the primary 22-month mission at Jupiter. In these ten orbits, *Galileo* will visit Europa, Ganymede, and Callisto at least twice at different viewing angles and typically at ranges of less than 600 miles (1,000 kilometers)—and in some cases at 120 miles (200 kilometers)! The closest approach that the *Voyagers* made to any Jovian satellite was 12,400 miles (20,000 kilometers). Thus *Galileo* promises a hundredfold improvement in the images of the satellite surfaces.

Galileo will complete its primary mission in October 1997—twenty years after the start of the project. Its scientific bounty will be tremendous.

---

toward the center of the solar system, others accelerated out of the solar system, while still others were flung far from the planets and the Sun yet still held feebly to the solar system by the Sun's gravity. The planets had been practicing gravity assist long before manmade space probes came along.

By spin and gravitational contraction, the solar nebula had become a flattened disk, but thick enough so that planetesimals and planets could approach one another for gravitational encounters at modest angles of inclination. The cometary planetesimals could pass above or below the poles of the protoplanets (like *Voyager 2* at Neptune) and thus change not only their speed and direction but their inclination as well. Comets were sprayed out from the protoplanets in every conceivable direction.

It was a titanic snowball fight with the Sun and planets throwing rock-flecked balls of ice in all directions—up, down, and toward one another. The supply of icy ammunition was enormous. Trillions of comets were heaved out of the solar system and lost forever. Trillions of others disappeared from existence in collisions with the large bodies. Still others, again numbering in the trillions, were flung far from the Sun but not beyond its gravitational hold.[1] With all their different distances and orbital inclinations, these discarded building blocks of the planets took the form of a vast, deep, spherical, low-density cloud of primordial material, each a time capsule preserving a precious record of the solar system when it was young. The Oort Cloud of comets had formed.

Stars and interstellar clouds of gas and dust passing close by the solar system from time to time would disturb the fragile orbits of these sen-

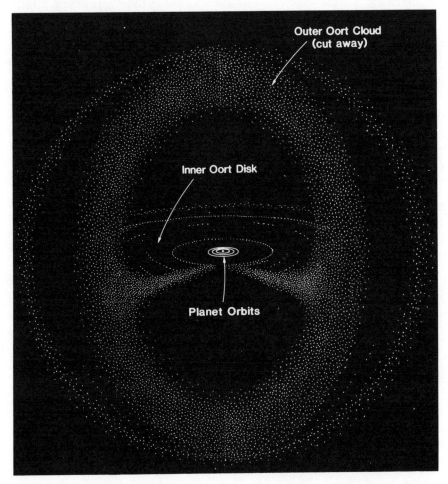

The Oort Cloud of comets, composed of an inner ring beginning not far beyond Pluto and broadening into a spherical outer cloud that extends almost halfway to the nearest star. Passing stars and nebulae strip away the outer comets, but disturb the inner disk enough so that a new supply of comets is ejected outward.

tinels and scatter them outward and inward—millions lost to interstellar space on each occasion, other millions dropping sunward to warm and shed their gas and dust and shine briefly in the sunlight. After 4.5 billion years, a creature would arise on the third planet from the Sun and call these apparitions comets, eventually to recognize from their displays and the material they shed that they were emissaries from far away and long ago—and that they preserved the chemical and physical secrets the creature sought about the origin of the solar system and the conditions from which life had emerged. We have survived, the comets said. Look around the solar system and see what we have built. Come with us out far from the Sun and see how we did it. Can you make that journey? If you can, you will be amazed.

Planets of Inclination

As the interplanetary snowball fight began to subside, two of the three outermost worlds were lying on their sides. In their formation, all the planets had withstood the crashes of millions of planetesimals—not always so small and not always so gentle. Some of these impacts, occurring near the poles, had tended to knock these planets on their sides.

The process was not quite like a bowling ball knocking bowling pins on their sides. Unlike bowling pins, Uranus and Pluto were spinning like gigantic tops or gyroscopes, which gave them considerable angular momentum. Give the top of a spinning gyroscope a gentle poke from the side and it does not fall over. Instead it precesses to a new axial orientation in a period of time matching the duration of the poke.

Except where other factors intervened, the final tilt of each planet was accidental—the sum of where and when and how hard each planet was struck. The tidal forces of the Sun compelled Mercury and Venus to stand upright as they spun. The gravitational collapse of solar nebula gases onto the rocky cores of the massive protoJupiter and protoSaturn to make them gas giants tended to overwhelm the the effect of planetesimal collisions. For the remaining planets, their axial inclinations are random. The tilts of Earth, Mars, and Neptune are moderate. For Uranus and Pluto, impacts have resulted in capsized planets.

But regardless of the mechanism that twists planets on their sides, it is clear that they developed that orientation before they had completed their accretion. Their satellites tell us that. Uranus and Pluto may have oddly oriented axes of rotation, but the orbits of their moons are entirely normal. The moons revolve around Uranus and Pluto in the same direction that the planets rotate. And they revolve around their planet's equator.

Satellites form around large planets the same way that planets form around a star—accreting from a dusty nebula. Satellites and planets start to form at the same time, but the smaller bodies form more slowly. If

Uranus assumed its reclined attitude by impact, it must have happened so early that the present moons had not yet begun to form. For Uranus, after its formation was complete, to have captured 15 moons all orbiting above its equator and all revolving in the same direction would strain credulity.

Pluto, with its weak gravity, could not have captured a comparatively large moon such as Charon. With its feeble gravity, it could not have had a moon-forming nebula around it. Charon more likely originated from an impact that shattered the original Pluto. The fragments reaccreted into a pair of bodies locked to one another by gravity.

But what if Uranus, Pluto, and their satellites had completed or nearly completed their formation with their rotation axes upright and then were knocked on their sides by impact? After the planets and asteroids took shape, there were still enough large and small planetesimals roaming the solar system to knock a planet askew. But the pivoting of Uranus and Pluto to a new axial orientation would not have flipped the orbital plane of their satellites to match. There is nothing in the impact event that would cause the satellite orbits to precess immediately or even over an extended period of time so that they could resume their original orbital positions over their planet's equator and with their original orbital motions in the same direction as the planet was turning.

Thus, Uranus and Pluto must have assumed their tilted rotations so early in their formations that their moons formed with their orbits tilted to match their planets' equators. They have never known their planets with other than the inclined rotations that we see today.

Rare Survivors

The comets that formed in the planetary realm of the solar system are all gone from their birthplace now. With uncountable trillions of comets filling the disk from which the Sun and planets would form, it must have seemed like an inexhaustible supply. But the success of the grains of dust and the crystals of ice in accreting into tiny nuclei, in gathering by adhesion more material to grow to comet size, and then in growing larger still by gravity—that very success destroyed or banished them. From these tiny planetesimals came larger planetesimals, and from these planetesimals came the planets, each consuming a billion or a trillion or more comets in their growth; then each, by its gravity, scattering trillions more—comets lost to still more collisions or to speeds that carried them beyond the Sun's grasp.

For a while there must have been many large planetesimals, a midway step in forming planets and, if they remained today, potentially instructive in understanding the evolution of the solar system. But the demands of gravity are insatiable. One by one these building blocks of

A RENDEZVOUS WITH A COMET

by Dr. Donald K. Yeomans, Jet Propulsion Laboratory and
Astrometry Team Leader, International Halley Watch

Since the beginning of recorded history, comets have been feared as agents of destruction—fireballs thrown at a sinful Earth from the right hand of an avenging god. Ancient cultures as diverse as the Greeks, Chinese, and Maya fearfully watched and wondered at the strange appearances of comets in an otherwise orderly sky.

Comets are now thought to be the initial building blocks that combined to form the planets in the outer solar system. In the 4.5 billion years since then, these remnants from the planet-building process have spent the vast majority of their lifetimes far from the Sun. Hence the ices that make up com-

ets have been retained in a nearly unaltered state and still offer clues to the original chemical composition of the outer planets.

The small, rocky asteroids, most of which orbit the Sun between Mars and Jupiter, are also expected to consist of preserved primitive matter typical of the region in which they formed.

Together, comets and asteroids provide the opportunity to study the building blocks of both the inner and outer planets of our solar system.

Despite recent advances in cometary research, owing in large part to the fleet of five international spacecraft that flew past Comet Halley in March

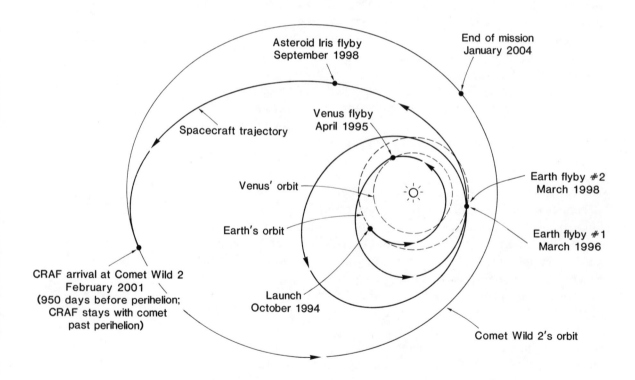

Comet Rendezvous Asteroid Flyby (CRAF)
Mission—1994 launch for Comet Wild 2 and
Asteroid Iris

1988, there remain many unanswered fundamental questions concerning comets.

- What is the nature of the jet black surface that was observed on Comet Halley, and what is the composition of the icy material below the comet's surface?
- How do comets evolve with time, and where and under what conditions did the cometary ices form?
- How did the planetary formation processes in the inner solar system differ from those in the outer regions?

These questions, and many more, could be answered with a spacecraft mission that has been proposed by NASA to rendezvous with a comet.

Upon arrival, the spacecraft would scrutinize the comet as it moves toward the Sun and evolves from a quiescent body to an active comet whose ices vaporize and throw off gas and dust into its enormous atmosphere.

One of the mission options under study would launch a Mariner Mark II spacecraft in October 1994, allowing the intensive study of short-period Comet Wild 2 to begin in February 2001. En route to the comet, the heavily instrumented spacecraft would fly closely past Iris, one of the larger asteroids. The spacecraft's cameras, as well as a host of other instruments, would examine the asteroid and, later, the comet itself.

Upon arriving at the comet, the spacecraft would be maneuvered into a path that matches that of the comet. After a brief reconnaissance to determine the comet's mass and characterize its size, shape, and surface features, the spacecraft would orbit the comet for several months.

During this time, the spacecraft's cameras and infrared detectors would map the comet's surface and note the differences in its thermal properties. The tiny dust grains in the comet's atmosphere would be subjected to chemical analysis and examined under a miniature electron microscope within the spacecraft. The chemical composition and magnetic properties of the comet's gases could be determined with a variety of instruments.

But potentially the most important instrument package will be shot, like a dart, just below the comet's surface. This penetrator device would

CRAF (Comet Rendezvous Asteroid Flyby) spacecraft design
Jet Propulsion Laboratory

CRAF spacecraft near a comet as its surface ices begin to vaporize
NASA/Jet Propulsion Laboratory artwork

measure the physical, chemical, and thermal properties of the subsurface ices, giving a first look at the primitive mixture from which our solar system formed some 4.5 billion years ago.

For several millennia, mankind's fear of a large comet striking the Earth was nearly universal. It seems ironic that by the beginning of the next millennium, mankind may strike a comet first—to take a peek at the conditions under which our solar system formed.

---

planets vanished—in collisions with the Sun or planets, in close encounters with the large planets that hurled them out of the solar system never to return or, perhaps, propelled them to new orbits at great distance from the Sun so that they have not yet been detected.

In the planetary realm, few large planetesimals are left. Perhaps Pluto is one. And Charon.[2] Rare survivors, preserved by the same gravitational forces that destroyed or expelled all the others. Almost as if prodigal gravity, having stripped the environment of a record of its past, decided to set up a wilderness area as a kind of natural park of the solar system to preserve something of special value in danger of disappearance.

But in truth it was no accident or last-minute salvation. The same inexorable law of gravity that sent every other planetesimal to its doom or exile also governed and preserved the orbit of Pluto and Charon. The orbital harmonics of Jupiter, Saturn, Uranus, and Neptune, which had destroyed, captured, or banished all the other planetesimals, provided a safe haven for one pair. Pluto and Charon formed where they did and survived in their peculiar orbit because it was a zone protected by gravity. Each planet alone might have diverted Pluto and Charon to oblivion, but with each adding to or balancing the other's gravitational nudges, the result for Pluto and Charon was a stable orbit over the duration of the age of the solar system.

The Peculiar Satellites of Neptune

Triton, however, lacked a protective resonance in its orbit. Gradually, over millions of years, it drew closer and closer to Neptune. Both were orbiting the Sun in the same direction. Both were orbiting at nearly the same speed. About 4.5 billion years ago, Triton passed too close, catching up with Neptune from behind. With both traveling in the same direction at nearly the same speed, the rate of closure was slow. Then faster, as Triton was accelerating under the influence of Neptune's enormous gravity. Now so close that Neptune's gravitational influence on Triton was much greater than that of the distant Sun.

Triton's path was altering rapidly. It passed sunward of Neptune very close and then veered around in front of Neptune, caught in its gravity. Its speed carried it outward far from Neptune but only to a certain distance. Back it fell along an elliptical path toward its new master, sweeping in close and around, then far out again, orbiting Neptune in the opposite direction it had been circling the Sun.

It was a highly erratic orbit and unstable. Already forces were remolding it. In the early days of the solar system, each large planet possessed its own accretion disk of gas and dust from which it and its satellites had formed—a miniature solar system with a planet, not a star, at its center. When Triton was captured, Neptune was losing its primordial disk but there were still enough particles near the planet to exert a substantial drag on Triton every time it passed close. At each periapsis, when Triton was traveling fastest, the debris around Neptune slowed its speed so that its momentum could not carry it outward to the distance it had gone just one orbit earlier. The far point in Triton's orbit was shrinking; the orbit was becoming rounder.

Even when the primordial nebula was gone, the orbit of Triton continued to be modified and rounded. Neptune created a tidal bulge in Triton and compelled that bulge to stay pointed in its direction. At its closest approach to Neptune, tidal forces on Triton's bulge decelerated the new satellite, preventing it from traveling as far from Neptune. When Triton was farthest from Neptune, the planet's tidal forces accelerated the moon so that it would not fall in as close. Triton's orbit became more and more circular.

Neptune had gained a moon. A big moon. (Later, it would capture a second moon, known today as Nereid.) But Neptune paid a price. The spectacular capture of Triton was a wild gravitational adventure that cost Neptune all of its original large moons. Some must have hit Triton as Neptune's gravity dragged it repeatedly across their orbits. Others must have been accelerated out of the Neptune system. We see 16 moons for Jupiter, 17 for Saturn, 15 for Uranus—but only 2 for Neptune. *Voyager*

FOUR YEARS WITH THE RING-MASTER: THE CASSINI MISSION TO SATURN

Three NASA spacecraft, *Pioneer 11* and the *Voyagers,* have flown by Saturn, giving scientists surprises everywhere they looked—the atmosphere (jet-stream winds of 1,000 miles per hour), the rings (thousands of them), the modest-sized moons (radically different histories, to judge by their surfaces), and Titan, the largest moon (a nitrogen atmosphere with 60 percent greater surface pressure than the Earth's).

Imagine what a spacecraft could do if it went into orbit around Saturn and could examine the planet, its rings, and its moons for at least four years, using a host of advanced instruments, including CCD cameras with ten times better resolution than the cameras aboard the *Voyagers.* Imagine that this orbiter carried a special probe craft that would descend into the atmosphere of Titan and perhaps survive to see the surface, which is shrouded from view on Earth by constantly hazy, orange-colored skies.

That mission is *Cassini,* named in honor of (not the modern fashion designer) but the seventeenth-century Italian-born French astronomer and telescope pioneer who discovered four moons of Saturn and a 3,000-mile (5,000-kilometer) gap in the ring system (the Cassini Division), and who proposed that the rings are composed of particles that orbit Saturn like miniature moons.

But the *Cassini* mission remains only under study as NASA's planetary science budget remains at low ebb. From 1962 to 1978, the United States

launched 27 successful unmanned missions to the Moon and planets. Since 1978, the United States has launched none. NASA flights to Venus (*Magellan*), Jupiter (*Galileo*), and Mars (*Mars Observer*) are awaiting the return of the Space Shuttle. *Magellan* and *Galileo* are scheduled for launch in 1989, *Mars Observer* in 1992.

American planetary space research once led the world. Now, despite a quarter century of magnificent success, no plan for the future has been adopted. Beyond completed but stranded spacecraft waiting for a chance to fly, planetary research in the American space program is close to extinction.

But if Congress provides the funds, *Cassini* could be launched in 1995, soon after the Comet Rendezvous Asteroid Flyby (CRAF) mission. Like CRAF, *Cassini* would be built around the Mariner Mark II, a flexible, multi-purpose spacecraft that would be assembled on a production-line basis to save costs. To this basic craft would be added standard and special components needed for specific jobs. ESA, the European Space Agency, would contribute the *Titan* probe.

Originally, *Cassini* was conceived for launch by Shuttle into Earth orbit, where an attached Centaur upper stage would boost it directly to Saturn, but safety concerns led to the cancellation of the powerful Centaur as part of a Shuttle payload.

Cassini must now rely on a Titan IV/Centaur expendable launch vehicle. But this two-stage rocket

2 will almost certainly find more, but they will be close to Neptune, will not be large, and may not be numerous. Triton saw to that.

And Triton paid a price, too. It must have taken a fierce pounding from Neptune's original satellites as some vanished from existence by their impacts on the usurper. Some may well have been the size of Miranda or Charon or even Oberon and Titania. The collisions shook the small

Cassini spacecraft design
Jet Propulsion Laboratory

does not have enough thrust to send the spacecraft directly to Saturn, so *Cassini*, like *Galileo* and CRAF, will rely on gravity assists. Its booster will send it first part way to Jupiter, where it will be decelerated so that it will fall back inside the orbit of Earth, to 0.9 astronomical units. As its momentum begins to carry it outward again, it will pass the Earth and receive a gravitational assist toward Jupiter. A second assist at Jupiter will send *Cassini* on to Saturn, for arrival in 2002.

But the spacecraft will not have wasted its circuitous 6.8 years en route. On its initial outbound journey into the asteroid belt, it will be targeted to fly by one asteroid. On its journey to Jupiter, it will encounter another. And, although it will only fly through the Jovian system, its advanced instrumentation should provide important information about Jupiter after *Galileo* has completed its service.

And when *Cassini* arrives at Saturn, it will take up orbital residence there at about the same distance from the ring-master as the Moon is from Earth for a research career that is expected to last at least four years.

With *Galileo*, CRAF, and *Cassini*, the era of long-duration, close-up study of the distant bodies in our solar system will have begun.

planet-now-moon, melting its surface, with heat and vibrations penetrating deep into the interior.

Less spectacular but even more powerful in reshaping Triton were the tidal forces of Neptune, pulling the captured body out of shape so that its interior was heated by pressure and friction past the melting point. Heavy elements sank to the core. Light elements rose toward the surface. The imprisoned Triton differentiated in what must have been a frenzy of tectonic upheavals and volcanic eruptions. Gases billowed above the landscape.

Based on more ordinary satellites close to other giant planets, we would expect to see on Triton, if its atmosphere is transparent, a tortured crust that indicates an era of brief, intense geologic activity soon after the moon was formed. But Triton should go beyond the usual if it endured the gravitational ordeal of capture and millennia of tidal strain that bent its path toward circular. It has changed since it was an unattached planetesimal like Pluto.

What story will it tell? That of a nearly pristine remnant of the formative days of our solar system, the closest approximation to a large planetesimal we can examine until we send a spacecraft to Pluto? Or has Triton changed so much through its violent history that the face it shows and the story it tells will be of the ordeals it has endured?

On August 25, 1989, Triton will at last have a visitor. And *Voyager 2* will send answers—and new mysteries—across 2.7 billion miles (4.3 billion kilometers) of space to the curious inhabitants of planet Earth.

Endnotes

CHAPTER 1. THE DISCOVERY OF URANUS

1. The often-told story that William essentially deserted the army and was later pardoned by King George III during their first meeting is not correct. As a bandsman, William was not automatically a member of the army and, because of his youth, had never been officially sworn in. However, the official discharge (signed in 1762) was slow in coming, and this may have been the reason that William did not return to Hanover before it arrived in 1763. It then took him more than a year to save sufficient money for the visit.

2. Told by a friend named Dr. Miller, as quoted by J. L. E. Dreyer in his biographical account of Herschel in William Herschel, *The Scientific Papers*, edited by J. L. E. Dreyer (London: Royal Society and Royal Astronomical Society, 1912), vol. 1, p. xx:

   Having afterwards asked Mr. H. by what means, in the beginning of his performance, he produced so uncommon an effect, he replied, "I told you fingers would not do!" and producing two pieces of lead from his waistcoat pocket, "One of these," said he, "I placed on the lowest key of the organ, and the other upon the octave above; thus, by accommodating the harmony, I produced the effect of four hands instead of two."

3. The problem of chromatic aberration was lessened by the use of compound lenses, introduced by Chester Moor Hall in 1729 and popularized by John Dollond and his son Peter in the 1850s. But telescope makers had great difficulties in obtaining glass of good quality for even small compound objective lenses and very rare success in making them larger than 4 inches. Nevil Maskelyne, the astronomer royal, purchased a 3.75-inch Dollond achromatic refractor for the Royal Observatory at Greenwich.

4. Constance A. Lubbock, ed., *The Herschel Chronicle* (Cambridge: Cambridge University Press, 1933), p. 145. Lubbock was Herschel's granddaughter.

5. William Herschel, *The Scientific Papers*, vol. 1, p. xxix.

6. Lubbock, pp. 79–80.

7. It was known at the time, however, that Halley's Comet and several others were members of the solar system, although their orbits stretched out beyond Saturn and even Uranus.

8. Letter of June 3, 1782, in Mary Cornwallis (Mrs. John) Herschel, *Memoir and Correspondence of Caroline Herschel* (New York: D. Appleton, 1876), p. 47; also Lubbock, p. 115.

9. Undated letter, in Lubbock, p. 117. Aubert's finest instrument was a 6-inch Cassegrainian reflector made by famed optician James Short. Its short focal-length/wide-field design (f/4) gave it a truncated, fat look and earned it the nickname "Short's Dumpy."

10. Herschel wrote this in response to Dr. Hutton's request for a brief sketch of his life. See Lubbock, pp. 78–79.

11. Lubbock. p. 95.

12. Herschel received no official title such as king's astronomer or private astronomer to the king.

13. A. F. O'D. Alexander, *The Planet Uranus: A History of Observation, Theory and Discovery* (New York: American Elsevier, 1965), p. 54.

14. William Graves Hoyt, *Planets X and Pluto* (Tucson: University of Arizona Press, 1980), p. 25.

15. According to a letter to Caroline on July 3, 1782, Herschel had first used this simulation that evening for the princesses and other ladies of the court. This was before the king granted him a subsidy for his astronomical research. "The effect was fine and so natural," wrote Herschel, "that the best astronomer might have been deceived." Lubbock, p. 118.

16. "Herschel and the Construction of the Heavens" in *Uranus and the Outer Planets*, edited by Garry Hunt

(Cambridge: Cambridge University Press, 1982), p. 55.

CHAPTER 2. THE FERVOR FOR NEW PLANETS

1. Bonnet's *Contemplation de la nature* was first published in 1764. Some accounts of Bode's Law still report that in 1741 the German astronomer Christian Wolff developed this progression and suggested that perhaps a planet was missing between Mars and Jupiter. In 1783, Titius incorrectly (for reasons unknown) credited Wolff as a progenitor of the relationship. However, Stanley L. Jaki found that Wolff had developed only a sequence of numbers to help students remember relative planet distances. He never suggested a mathematical progression or a gap requiring a missing body. Stanley L. Jaki; "The Early History of the Titius-Bode Law," *American Journal of Physics* 40 (July 1972): 1014–1023.

2. In the second (1772) and subsequent editions, this editorial by Titius appeared instead as a footnote signed "T."

3. In Titius' formulation, there was no division by ten to make the sequence correspond to astronomical units.

4. My phrasing. A history of Bode's Law is presented by Stanley L. Jaki in "The Early History of the Titius-Bode Law," *American Journal of Physics* 40 (July 1972): 1014–1023; "The Original Formulation of the Titius-Bode Law," *Journal of the History of Astronomy* 3 (1972): 136–138; and "The Titius-Bode Law: A Strange Bicentenary," *Sky and Telescope* 43 (May 1972): 280–281.

5. *Anleitung zur Kenntniss des gestirnten Himmels* (Introduction to the Study of the Starry Heavens), first published in 1768 with a slightly different title.

6. In 1801, Bode said he got the relationship from Titius' second translated edition of Bonnet (1772), but, as Jaki has pointed out, that was not possible because Bode's book was published in January 1772, prior to the second edition of Titius' translation, in which his contribution was now a footnote signed "T."

7. J. E. Bode, *Deutliche Anleitung zur Kenntniss des gestirnten Himmels*, 2d ed. (Hamburg: D. A. Harmsen, 1772), p. 462. Also Jaki, "The Early History of the Titius-Bode Law," 1015.

8. This was the second meeting on the subject that Zach had arranged. The first was at his observatory in Gotha, Germany, in 1796.

9. Zach, Schröter, Olbers, Harding, and probably Ferdinand Adolph von Ende and Johann Gildemeister, according to Martin Grosser, *The Discovery of Neptune* (Cambridge: Harvard University Press, 1962), p. 30. Robert W. Smith says Bode was in attendance; see "The Impact on Astronomy of the Discovery of Uranus," in *Uranus and the Outer Planets,* edited by Garry Hunt (Cambridge: Cambridge University Press, 1982), p. 83.

10. Article on Piazzi by Giorgio Abetti in *Dictionary of Scientific Biography* (New York: Charles Scribner's Sons, 1974).

11. Zach had located a suspicious object on December 7, 1801, but had been unable to confirm it as the elusive body because of bad weather.

12. Robert W. Smith cites an unsigned 1802 article in *Annelen der Physik.* See "The Impact on Astronomy of the Discovery of Uranus" in *Uranus and the Outer Planets,* p. 85.

13. Bode's Law may make a comeback of sorts through the work of Jaume Llibre, a mathematician at the Autonomous University of Barcelona in Spain. He is exploring how the mass of the Milky Way, concentrated toward the center of our galaxy, could have affected where planets formed around our Sun.

    In the United States, Edward Belbruno, at the Jet Propulsion Laboratory, is applying Llibre's work to the planets and their satellite systems. In part, his efforts could lead to an understanding of the probabilities and processes of satellite captures and to positions where spacecraft could take up orbit around a moon or planet with the least expenditure of energy. (Edward Belbruno, personal communication, March 4, 1988.)

CHAPTER 3. TROUBLE WITH URANUS

1. Herschel's work on double stars in 1802 and 1803 showed that many were binary stars revolving around each other and not accidents of line of sight. He and most other scientists were confident that gravity held these systems together, but the uncertainty about distance to the stars at that time made proof of gravity's role at stellar distances impossible.

2. Most notably Placidus Fixlmillner in 1784 and Jean Baptiste Joseph Delambre in 1788. Fixlmillner was the first to point out that the prediscovery and postdiscovery observations of Uranus could not be reconciled. Delambre was the first to incorporate the gravitational perturbations of Jupiter and Saturn on Uranus in his calculations.

3. As seen from Earth, the brightness of Uranus doesn't change much as it revolves around the Sun because Earth's orbit is comparatively tiny and thus its distance from Uranus is always large. The average distance of the Earth from Uranus (which is the same as the Sun-Uranus distance) is 1.8 billion miles (2.9 billion kilometers). At its closest the Earth is 1.7 billion miles from Uranus; at its farthest, 1.9 billion miles. The difference between closest and farthest is about 11 percent, resulting in a difference in brightness of about 20 percent. The phase of Uranus is not a significant factor in its brightness because its orbit is so far outside the orbit of the Earth that Uranus always presents a full or very nearly full phase. At its faintest Uranus is about magnitude 5.8.

4. Bode was assisted in this identification by Placidus Fixlmillner.

5. Le Monnier publicly acknowledged 4, but Dennis Rawlins examined Le Monnier's observation journal and found that Le Monnier had penned marginalia indicating that he recognized he had observed Uranus 9 times without recognizing it as something other than a star. See Dennis Rawlins, "The Unslandering of Sloppy Pierre," *Astronomy* 9 (September 1981): 24, 26, 28.

   Did Le Monnier publish only his earliest observations because he was embarrassed by his oversight, or as science historian Ruth Freitag suggests, was he more concerned about the usefulness of his positions, feeling that the others, so closely spaced, would not help much in the effort to refined orbit calculations? (Personal communication, March 24, 1988.)

   Le Monnier actually recorded but did not recognize Uranus 12 times.

6. Twenty-two prediscovery sightings of Uranus are now known.

7. Rawlins.

8. Alexis Bouvard, *Tables astronomiques publiées par le Bureau des longitudes de France contenant les tables de Jupiter, de Saturne et d'Uranus, construites d'après la théorie de la mécanique céleste* (Paris: Bachelier et Huzard, 1821), p. xiv. Cited by Morton Grosser, *The Discovery of Neptune* (Cambridge: Harvard University Press, 1962), p. 42.

9. Alexis Claude Clairaut, *Journal des sçavans* (Paris, 1759), p. 86. Cited by Grosser, p. 49.

10. George B. Airy, "Account of Some Circumstances Historically Connected with the Discovery of the Planet Exterior to Uranus," *Monthly Notices* of the Royal Astronomical Society 7 (1846): 124–125.

11. Valz, letter on Halley's Comet, *Comptes rendus* 1 (September 21, 1835): 130. Cited by Grosser, p. 51.

    The problem puzzling Valz was the discrepancy between the carefully calculated perihelion date and the actual date of the comet's passage. This discrepancy is now known to be caused by non-gravitational forces—vaporization from the surface of a rotating comet that either brakes or accelerates the body and thus alters its orbit slightly.

12. *Astronomische Nachrichten* 13 (1835): col. 94. Cited by Grosser, p. 51.

13. Cited by Grosser, pp. 51–52.

14. Mädler, *Populäre Astronomie* (Berlin: C. Heymann, 1841), p. 13. Cited by Grosser, p. 56.

CHAPTER 4. NEPTUNE: THE PLANET FOUND ON A SHEET OF PAPER

1. W. M. Smart, *John Couch Adams and the Discovery of Neptune, Occasional Notes of the Royal Astronomical Society,* No. 11 (August 1947), p. 42. A poster bearing these words dates from 1847. (Smart was the first John Couch Adams Professor of Astronomy at Cambridge University.)

2. Smart, p. 45.

3. Library of St. John's College, Cambridge University.

4. George Biddell Airy, "Account of Some Circumstances Historically Connected with the Discovery of the Planet Exterior to Uranus," *Monthly Notices* of the Royal Astronomical Society 7 (1846): 145.

5. In the case of Saturn and Uranus, Bode's Law worked well but seemed slightly on the long side. Perhaps Adams was compensating for that.

6. George Biddell Airy, *Autobiography,* edited by Wilfrid Airy (Cambridge: At the University Press, 1896), p. 172.

7. A year later, when the consequences had been played out, Adams and Airy were comparing recollections of events by mail. In a letter dated November 18, 1846, Adams noted: "I was . . . much pained at not having been able to see you when I called at the Royal Observatory the second time, as I felt that the whole matter might be better explained by half an hour's conversation than by several letters." (Smart, p. 56.)

8. This assessment is well expressed by Morton Grosser in *The Discovery of Neptune* (Cambridge: Cambridge University Press, 1962), p. 91.

9. This story is based on Lassell's recollection as told to astronomer Edward S. Holden in 1876. Holden did not publish this account until 1892, after the deaths

of Lassell and Adams. However, Robert W. Smith believes that Lassell's memory was faulty and that these events occurred in September 1846, only slightly before Neptune was found at the Berlin Observatory. In any case, Adams' mathematical results were not known to a wide circle of British scientists before the optical discovery of Neptune. "William Lassell and the Discovery of Neptune," *Journal for the History of Astronomy* 14 (February 1983): 30–32.

10. Asked later why he didn't immediately respond to Airy's question, Adams replied, "I should have done so; but the enquiry seemed to me to be trivial." (Smart, p. 57.)

11. Harold Spencer Jones, *John Couch Adams and the Discovery of Neptune* (Cambridge: Cambridge University Press, 1947), p. 22.

12. *Comptes rendus* 22 (June 1, 1846): 918.

13. This and the following quotations from Airy's correspondence may be found in Airy, "Account," pp. 132–136.

14. Smart, p. 59.

15. John Herschel, "Le Verrier's Planet," *Athenaeum* (October 3, 1846): 1019.

16. Herbert Hall Turner, "Obituary of Johann Gottfried Galle," *Monthly Notices* of the Royal Astronomical Society 71 (February 1911): 279.

17. Willy Ley, *Watchers of the Skies* (New York: Viking Press, 1963), p. 411.

18. Smart suggests that Encke may have been waiting for the next chart to be printed so that he could economize by sending two at once (p. 80).

19. In celestial geocentric longitude, Le Verrier's prediction for the evening of September 23, 1846, was 324 degrees 58 minutes. Galle found it at 325 degrees 52 minutes 45 seconds—a difference of about 55 minutes of arc—less than one degree.

20. *Comptes rendus* 23 (October 5, 1846): 659.

21. The exact date is not certain, but Challis wrote Arago on October 5, 1846, to say that he had seen a disk on September 29, before news of the planet's discovery reached him. If Kingsley's account is accurate, the detection of a disk may have taken place during the second week in September, but that would presume that Challis suppressed the earlier observation.

22. Airy, "Account," p. 143.

23. Grosser, p. 123.

24. Grosser, p. 126.

25. Grosser, p. 120.

26. Smart, p. 65.

27. Smart, p. 65.

28. François Arago, "Examen des remarques critiques et des questions de priorité que la découverte de M. Le Verrier a soulevées," *Comptes rendus* 23 (October 19, 1846): 741–754.

29. Grosser, p. 133.

30. Adams' full work would normally have been published by the Royal Astronomical Society, but it was busy at the time publishing a long paper by Airy on the longitude of Valentia. William Samuel Stratford, superintendent of the *Nautical Almanac*, was so impressed by Adams' accomplishment that he intervened and printed Adams' research as an appendix to the *Nautical Almanac* for 1851 (published late in 1846).

The complete title of Adams' report was "An Explanation of the Observed Irregularities in the Motion of Uranus, on the Hypothesis of Disturbances Caused by a More Distant Planet; with a Determination of the Mass, Orbit and Position of the Disturbing Body."

31. *Monthly Notices* of the Royal Astronomical Society 7 (November 13, 1846): 150.

32. Smart, p. 83.

33. Olin J. Eggen, "James Challis," *Dictionary of Scientific Biography* (New York: Charles Scribner's Sons, 1974).

34. N. T. Roseveare, *Mercury's Perihelion from Le Verrier to Einstein* (Oxford: Clarendon Press, 1982).

35. Airy wanted to add this responsibility to his duties as astronomer royal for extra salary, but the admiralty was not willing. Two of Airy's chief assistants applied unsuccessfully for the post. Airy expressed great satisfaction that it was given to John Russell Hind. See Airy, *Autobiography*, p. 216.

CHAPTER 5. PERCIVAL LOWELL AND PLANET X

1. Among them was U.S. Naval Observatory astronomer David Peck Todd, who in 1877 used a graphical analysis of the perturbations of Uranus to predict a planet at 52 astronomical units with a period of 375 years, a diameter of 50,000 miles, an apparent disk of about 2 arc seconds, and a magnitude of 13 or fainter. Using the Naval Observatory 26-inch refractor, he searched for the planet for 30 nights from November 1877 to March 1878.

Other early seekers for a trans-Neptunian planet included Jacques Babinet (France, 1848), George

Forbes (Scotland), Camille Flammarion (France), Karel Zenger (Czechoslovakia), Oskar Reichenbach (Germany), Hans Emil Lau (Denmark), Gabriel Dallet (France), Theodore Grigull (France), Alexander Garnowsky (Russia), Choren M. Sinan (Turkey), and Thomas Jefferson Jackson See (United States). In 1907, Jean Baptiste Aimable Gaillot of France examined the motions of Uranus and Neptune and could find no indication of perturbations by an outlying planet. But later he agreed with William H. Pickering's 1908 analysis and predicted two planets himself. Martin Grosser, "The Search for a Planet beyond Neptune," *Isis* 55 (August 1964): 163–183.

2. Percival Lowell, *Mars* (Boston: Houghton Mifflin, 1895), p. v.

3. William Lassell, discoverer of Neptune's moon Triton in 1846, became the first astronomer on record to take his telescope to a remote site for better observing when he moved his 24-inch reflector from England to Malta in 1852. In 1860 he erected a 48-inch reflector there. He continued his observations from Malta until 1864, when he returned to England with his telescopes.

4. Clyde W. Tombaugh and Patrick Moore, *Out of the Darkness: The Planet Pluto* (Harrisburg, Pennsylvania: Stackpole Books, 1980), p. 83.

5. William Graves Hoyt, *Planets X and Pluto* (Tucson: University of Arizona Press, 1980), p. 104.

6. The ecliptic is the plane of the Earth's orbit around the Sun. All the planets out through Neptune revolve around the Sun in orbits that lie close to the ecliptic, hence the ancient development of the concept of the zodiac, the band of the sky centered on the ecliptic beyond which the known planets never strayed. The ecliptic is inclined only 1.6 degrees to the "invariable plane," the mean plane of the solar system.

7. Percival Lowell, *Memoir on a Trans-Neptunian Planet* (Lynn, Massachusetts: Press of T. P. Nichols, 1915; Memoirs of the Lowell Observatory 1.1).

8. Hoyt, p. 109.

9. Lowell never explained why he assumed this particular distance. Apparently he felt that 47.5 astronomical units was the best fit for a hypothetical planet that (1) was suspected of influencing the orbits of a family of short-period comets, (2) satisfied a graphical plotting of the residuals of Uranus, and (3) had a period of revolution that was commensurate with (standing in an even number ratio to) the periods of Uranus and Neptune so that the planets

would regularly line up with one another at specific places in their orbits to create the most measurable perturbations.

10. With modifications, this was the instrument Tombaugh used to find Pluto.

11. This was his second collapse from "neurasthenia." His first was in May 1897, and he did not fully recover for three years.

12. This and the following Lowell quotations may be found in Hoyt, pp. 124–132.

13. In his later calculations Lowell experimented with different estimates of the distance of Planet X from the Sun. He obtained solutions that seemed to account more satisfactorily for the perturbations in the motion of Uranus and Neptune when the distance of Planet X was closer to 44 astronomical units.

Mathematical derivations of the position of an unseen planet near the plane of the solar system from its gravitational effect on another planet (such as Adams and Le Verrier achieved and Lowell was attempting) yield two simultaneous solutions about 180 degrees apart—nearly opposite one another in the sky. We experience this same problem with ocean tides. High tides occur about twice a day: once when the Moon is above our location and again later that day when the Moon is on the opposite side of the Earth from our location. We cannot tell simply from looking at the tide whether the Moon is above us or on the opposite side of the Earth.

Thus Lowell's trans-Neptunian planet appeared to him on paper as two Planets X—at least one of them spurious. Other criteria would then have to be used to eliminate one of the solutions.

CHAPTER 6. THE DISCOVERY OF PLUTO

1. For Tombaugh's own description of the Planet X search, see Clyde W. Tombaugh and Patrick Moore, *Out of the Darkness: The Planet Pluto* (Harrisburg, Pennsylvania: Stackpole Books, 1980), especially pp. 114–130.

2. Tombaugh, personal communication, February 18, 1988.

3. Tombaugh quoted by William Graves Hoyt in *Planets X and Pluto* (Tucson: University of Arizona Press, 1980), p. 188.

4. Hoyt, p. 192. Earl Slipher was out of town on February 18 and did not immediately know about the discovery.

5. This telescope was a 40-inch reflector until 1925. Clyde Tombaugh explains, "In parabolizing the figure of mirrors, it is often difficult to avoid a 'turned down' edge, and telescope makers will use a blank two or three inches larger than the contract diameter of aperture. Then they diaphragm out this narrow defective zone at the edge by a metal ring covering. When it was discovered that the figure was true to the very edge, they merely removed the ring mask in the mirror cell, yielding a 42-inch aperture." (Personal communication, February 18, 1988.)

6. Hoyt, p. 200.

7. Tombaugh and Moore, p. 136. The astronomer "whose abrasive personality had made him universally detested" is identified by Patrick Moore in "The Naming of Pluto," *Sky & Telescope* 68 (November 1984): 400–401.

8. Venetia Burney's suggestion was telegraphed to the Lowell Observatory by astronomer and family friend Herbert Hall Turner. For more details, see Moore, 400–401.

   Clyde Tombaugh recalls "that the Lowell Observatory staff was already considering the name Pluto before the public announcement of the discovery and before the first outside suggestion by Venetia Burney." (Personal communication, February 18, 1988.)

9. Annie J. Cannon, "William Henry Pickering: 1858–1938," *Science* 87 (February 25, 1938): 179–180.

10. See Hoyt and also Tombaugh and Moore for a recounting of this debate.

Chapter 7. Toward Uranus

1. James Elliot, "Uranus: The View from Earth," *Sky & Telescope* 70 (November 1985): 415–419.

2. Hydrogen in the atmosphere of Uranus was identified by Vesto M. Slipher at the Lowell Observatory in the period 1902–1904. Methane was identified by Rupert Wildt in 1932. Slipher also used spectrography in 1912 to determine the rotation period of Uranus.

3. The discovery is described by James Elliot and Richard Kerr in *Rings: Discoveries from Galileo to Voyager* (Cambridge: MIT Press, 1984).

4. Using an orbital transfer trajectory, which requires the least energy.

5. Flandro also discovered that Pluto could be reached by gravity assist from Jupiter, but the positions of the outer planets in the 1980s precluded a single trajectory to all of them.

6. There were monthlong launch windows in 1976, 1977, and 1978, but 1977 presented by far the best opportunity.

7. It still has not adopted one. For a proposal, please see the report of the National Commission on Space, *Pioneering the Space Frontier* (New York: Bantam Books, 1986), and Thomas R. McDonough, *Space: The Next Twenty-Five Years* (New York: John Wiley & Sons, 1987).

Chapter 8. The Grand Tour of *Voyager 2*

1. I first heard this kind of characterization of *Voyager 2* expressed during the Uranus encounter by Edward C. Stone, project scientist for the *Voyager* missions.

2. The scan platform pivots in azimuth and elevation. It was the azimuth motion that stuck.

3. The roll axis of the spacecraft is the direction that the high-gain antenna (the big dish) points. The cameras and most of the instruments are mounted on a boom that extends perpendicular to the roll axis. So rolling the spacecraft moves the cameras in azimuth. The ability of the cameras to tip up and down in elevation is unimpaired.

   Charles Kohlhase explains that rolling the entire spacecraft was the preferred mode for taking pictures and data at Uranus because spacecraft roll was smoother and could match any target motion rate. The scan platform could not slew as smoothly, and because it used stepping motors, it would not always match target motion speeds. (Personal communication, March 7, 1988.)

4. A few months after launch, *Voyager 1* experienced a similar problem but lost a much larger portion of its memory.

5. The time taken to broadcast a picture should not be confused with the time it takes that message to reach the Earth. Because of the 483 million miles (778 million kilometers) between Jupiter and Earth, it would take the picture message about 43 minutes of space travel at the speed of light to arrive on our planet.

Chapter 9. Triumph at Uranus

1. Andrew P. Ingersoll, "Uranus," *Scientific American* 255 (January 1987): 38–45.

2. The Sun could stand directly above the Uranian poles if the rotational axis of the planet lay exactly in its plane of revolution. However, the axis of Uranus lies about 8 degrees from the plane of revolution, so the Sun can never be seen vertically above any point within 8 degrees of the poles.

3. Uranus is considered to be a retrograde rotator, that is, spinning on its axis in the opposite direction of the majority of planets. As one looks down on the north pole of Earth, our planet spins counterclockwise. As one looks down on the north pole of Uranus, the planet spins clockwise. If one defines east as the direction to the right when one looks north, then Uranus rotates east to west. On Uranus, then, the Sun rises in the west and sets in the east.

4. Because of the off-center magnetic axis, the latitudes of the magnetic poles of Uranus are different. The "positive" (north magnetic) pole has a latitude of 15.2 degrees south of the equator, while the "negative" (south magnetic) pole has a latitude of 44.2 degrees north of the equator (using the International Astronomical Union's definition for rotational poles in which the sunlit pole that greeted *Voyager 2* was the south rotational pole).

5. The magnetic field of Earth is generated deep within our planet but not at its center, because the inner core is solid. Instead, the Earth's magnetic field is generated around the center in the liquid outer core.

6. During a magnetic field reversal on Earth, the strength of the field drops to near zero, leaving our planet without the protection of this magnetic shield to deflect high-speed subatomic particles from the Sun and deep space.

7. Jupiter and Saturn do have auroras as well.

8. Robert Hamilton Brown, personal communication, April 9, 1988.

9. The best current measurements of excess radiation from the giant planets are as follows:

Jupiter	1.7
Saturn	1.8
Uranus	less than 1.2
Neptune	2.0 to 2.5

Neptune generates only a little more internal energy than Uranus but receives much less sunlight, so its excess energy ratio is higher. Data supplied by Andrew P. Ingersoll, personal communication, February 26, 1988.

10. Andrew P. Ingersoll, "Jupiter and Saturn," *Scientific American* 245 (December 1981): 90–108. Why should this process be under way at Saturn but not Jupiter? Professor Ingersoll explains that metallic hydrogen and helium can be kept mixed only under high temperatures. The temperature inside Saturn has cooled sufficiently for the helium to begin to separate out. This process should eventually happen at Jupiter and may be beginning already. (Personal communication, February 26, 1988.)

11. Based on data assembled by James Elliot and his colleagues on the gravitational field of Uranus derived from perturbations of its rings, Vladimir Zharkov of the Soviet Union was the first to propose that the mantle and atmosphere of Uranus were mixed together rather than in distinct layers.

12. A character named Ariel also appears in Shakespeare's *The Tempest*.

13. The orbit of Miranda is just slightly inclined (4 degrees) to the planet's equator.

14. At Jupiter's distance from the Sun and closer, asteroids are a hazard too.

15. One of the *Voyager* geologists, Eugene M. Shoemaker, suggests that Titania was blasted apart by a large planetesimal, but the shattered pieces of Titania remained in orbit and reassembled themselves by gentle collisions and mutual gravitation, thereby erasing all traces of the early bombardment phase. Torrence V. Johnson, Robert Hamilton Brown, and Lawrennce A. Soderblom, "The Moons of Uranus," *Scientific American* 256 (April 1987): 48–60.

16. The *Voyager* scientists often refer to these grooved ovoids as *circi maximi* (plural of *circus maximus*), named after the chariot racetracks of ancient Rome.

For two of the three ovoids, the other dimension was unknown because they extended out of sight across the equator.

17. Robert Hamilton Brown, "Exploring the Uranian Satellites," *Planetary Report* 6 (November/December 1986): 4–7, 18; also, personal communication, April 1, 1988.

18. Lawrence Soderblom and Bradford A. Smith quoted in Jonathan Eberhart, "Mysteriously Muddled Miranda," *Science News* 129 (February 15, 1986): 103–04.

19. "Towards a Theory for the Uranian Rings," *Nature* 277 (January 11, 1979): 97–99.

20. One of the two rings that *Voyager 2* discovered is not extremely narrow. 1986U2R, the innermost ring, is very tenuous, relatively dusty, and has a width of 1,500 miles (2,500 kilometers).

21. Ernst J. Öpik had proposed a similar idea in 1932.
22. Estimated for comets with an absolute B magnitude brighter than 18 by B. A. Smith, L. A. Soderblom, et al. in *"Voyager 2* in the Uranian System: Imaging Science Results," *Science* 233 (July 4, 1986): 43–64.
23. Carolyn C. Porco, *"Voyager 2* and the Uranian Rings," *Planetary Report* 6 (November/December 1986): 11–13.

Chapter 10. Appointment with Neptune

1. In September 1985, *Voyager 2* had been scheduled to pass only 800 miles (1,300 kilometers) above Neptune's north polar cloud tops and only 6,200 miles (10,000 kilometers) from Triton. But concern about collisions with particles in Neptune's arc ring system as *Voyager 2* passed through the equatorial plane on approach to the planet caused mission scientists and designers to elect a more conservative path in January 1987.
2. This comparison to Earth twilight was provided by James A. DeYoung, Nautical Almanac Office, U.S. Naval Observatory. The light level is approximately equal to the Sun 2 degrees below the horizon on a clear day with brightness measured by reflection off a flat surface.
3. This work was done by Dale P. Cruikshank, Robert Hamilton Brown, and Michael J. S. Belton.
4. Bradford A. Smith, Harold J. Reitsema, and Stephen M. Larson in 1979; Richard J. Terrile and Bradford A. Smith in 1983, using a CCD with the 100-inch (2.5-meter) telescope at the Carnegie Institution of Washington's Las Campanas Observatory in Chile; and Heidi B. Hammel in 1986, using a CCD with the University of Hawaii's 88-inch (2.24-meter) telescope at the Mauna Kea Observatory.
5. Harold J. Reitsema, William B. Hubbard, Larry A. Lebofsky, and David J. Tholen, "Occultation by a Possible Third Satellite of Neptune," *Science* 215 (January 15, 1982): 289–291.
6. William B. Hubbard explains, "There were two computer printouts from the telescope. One recorded the maximum and minimum reading every 3.4 seconds. The event is dimly visible in this record. The other printout recorded 5-second averages of the data. The event is also faintly detectable in the record, in channel 2. But we weren't convinced until we saw the full data display, with one data point every 0.01 second." (Personal communication, March 8, 1988.)
7. W. B. Hubbard, A. Brahic, B. Sicardy, L.-R. Elicer, F.

Roques, and F. Vilas, "Occultation Detection of a Neptunian Ring-like Arc," *Nature* 319 (February 20, 1986): 636–640.
8. Jack J. Lissauer, "Shepherding Model for Neptune's Arc Ring," *Nature* 318 (December 12, 1985): 544–545, and Peter Goldreich, Scott Tremaine, and Nicole Borderies, "Toward a Theory for Neptune's Arc Rings," *Astronomical Journal* 92 (August 1986): 490–494.
9. Personal communication, March 18, 1988.
10. Of all the bodies in the solar system, only Earth, Titan, and presumably Triton have predominantly nitrogen atmospheres.
11. At its closest approach to Neptune, *Voyager 2* will be traveling 61,000 miles per hour (27.3 kilometers per second) with respect to the planet.

Chapter 11. Voyagers to the Stars

1. The first prediction that the heliopause would lie well beyond Neptune and Pluto rather than between Jupiter and Uranus was by T. R. McDonough and N. M. Brice, "The Termination of the Solar Wind (Solar Wind Termination Distance as Function of Flux, Velocity and Interstellar Hydrogen Density, Velocity and Magnetic Field Strength)," *Icarus* 15 (December 1971): 505–510.
2. *Voyager* flight and stellar-encounter data in this chapter were provided by Robert J. Cesarone of the Voyager Navigation Team at the Jet Propulsion Laboratory, personal communication, February 1988; and Robert J. Cesarone, Andrey B. Sergeyevsky, and Stuart J. Kerridge, "Prospects for the *Voyager* Extra-Planetary and Interstellar Mission," *Journal of the British Interplanetary Society* 37 (March 1984): 99–116.

   Cesarone cautions that the distances and radial velocities of nearby stars are not known with great accuracy, and therefore the times and distances of *Voyager* encounters with these and other stars are only approximate.
3. AC stands for Astrographic Catalog; +79 is the star's approximate declination (degrees north, in this case, of the celestial equator); and 3888 is the catalog entry for that star in that region.
4. The closest passage of Ross 248 to the Sun is 2.9 light-years, about 35,000 years from now. AC+79 3888 will pass 3.0 light-years from the Sun in 40,600 years.
5. Carl Sagan, *Murmurs of Earth: The Voyager Interstellar Record* (New York: Random House, 1978).

CHAPTER 12. THE SMALLEST PLANET

1. Specifically, Kepler's third law as modified by Newton.
2. In 1966, Pluto failed to occult a star along its path, thereby confirming that it was less than 4,200 miles (6,800 kilometers) in diameter.
3. Merle F. Walker and Robert Hardie, "A Photometric Determination of the Rotational Period of Pluto," *Publications of the Astronomical Society of the Pacific* 67 (1955): 224–231.
4. Dale P. Cruikshank, Carl B. Pilcher, and David Morrison.
5. Donald M. Hunten and Andrew J. Watson, "Stability of Pluto's Atmosphere," *Icarus* 51 (1982): 665–667.
6. Christy explains, "I had initiated the request to take these plates two years earlier. Because we were doing Neptune and Uranus, I thought we should complete the trio. The primary purpose for Pluto was orbital-position measurement, but because we were measuring the position of the moons of Uranus and Neptune, I had included in my rationale the potential for discovering a moon of Pluto via possible perturbations in Pluto's motion." (Personal communication, February 15, 1988.)
7. Christy thought that the light variation period and the moon's orbital period were identical and informed Robert S. Harrington of this likelihood. "I assumed," said Christy, "that the light period was caused by the moon, which was incorrect. Harrington quickly (I don't recall when) realized that Pluto and Charon were tidally coupled, thus the *rotation period* [my italics] was identical to the orbital period. This was a critical insight which made early interpretation of the physics correct in every respect." (Personal communication, February 15, 1988.)
8. Robert S. Harrington, personal communication, January 26, 1988.
9. Edith Hamilton says Charon's boat crossed Acheron, the river of woe, and Cocytus, the river of lamentation, where the two waters flowed together.
10. Harrington and his wife named their daughter, born soon after the discovery, Ann Charon Harrington.
11. The combined mass of the Pluto-Charon system is 0.0026 the mass of Earth. From this measurement Pluto's mass could be approximated from its estimated size and density. Pluto's mass is about 0.0022 the mass of Earth if Charon's mass is about 15 percent of Pluto's.

12. In 1973, Leif E. Andersson and John D. Fix suggested that Pluto's axis was steeply tilted because of changes in the amount of light reflected from Pluto as it rotated. "Pluto: New Photometry and a Determination of the Axis of Rotation," *Icarus* 20 (1973): 279–283.
13. As defined by the researchers in the field, in which the north pole is the one that spins counterclockwise as seen from above. This designation differs from the International Astronomical Union's convention, which defines a planet's north pole as the one north of the ecliptic. The latitude of Pluto directly beneath Tombaugh's gaze was roughly 60 degrees south. (Marc W. Buie, personal communication, March 8, 1988.)
14. William B. McKinnon cautions that another interpretation is at least as likely: that Pluto's surface may be reflective methane ice and the dark equatorial band may be either a coating deposited on top of the reflective surface (like that on Saturn's Iapetus?) or methane that, under ultraviolet light from the Sun has chemically reacted to form dark-colored hydrocarbon chains. If Pluto has a crust of soft methane ice, McKinnon notes, the craters may have sagged and disappeared. (Personal communication, April 29, 1988.)
15. From Pluto at perihelion, the Sun subtends approximately 65 seconds of arc—30 times less than on Earth. The disk of the Sun as seen from Pluto is about the same size as the disks of Venus (at its brightest) and Jupiter (at its closest) when seen from Earth. To the eye they appear starlike.
16. At aphelion in the year 2113, when Pluto is 49.3 astronomical units (4.6 billion miles; 7.4 billion kilometers) from the Sun, visitors on Pluto will see the Sun 2,400 times fainter than the Sun appears on Earth. Even so, it will still be 160 times brighter than Earthlings see the full Moon.
17. Presuming that we landed on a level plain. For comparison, a person on Earth would see a companion disappear over the horizon at a distance of 4.25 miles (6.75 kilometers). On the Moon, a companion would disappear at a distance of 2.25 miles (3.5 kilometers).
18. Unless some of the asteroids have satellites and they exhibit this behavior. No satellites for asteroids have been found.
19. If Charon had been blanketed by methane ice 14 miles (22 kilometers) deep at the beginning of the solar system, all that ice would have vaporized by now. Donald M. Hunten and Andrew J. Watson,

"Stability of Pluto's Atmosphere," *Icarus* 5 (1982): 665–667.

20. Charon's altitude and azimuth in the sky could vary slightly according to the eccentricity and inclination of its orbit, but in a system where both bodies are close together and relatively close to the same mass, gravity compels the eccentricity and the orbital inclination to be zero or very nearly zero.

21. For this eclipse, Charon is imagined to be spherical.

22. August 25 according to universal time (at the prime meridian at Greenwich, England) but prior to midnight on August 24 at the Jet Propulsion Laboratory, mission control for *Voyager 2*, which is in the Pacific Time Zone.

23. "Professor Yamamoto's Suggestion on the Origin of Pluto," *Bulletin* of the Kwasan Observatory (Kyoto Imperial University) 3.288 (August 20, 1934).

24. Gerard P. Kuiper, "Further Studies on the Origin of Pluto," *Astrophysical Journal* 125 (January 1957): 287–289; and "The Formation of the Planets," *Journal of the Royal Astronomical Society of Canada* 50 (1956): 171–173.

25. Neptune and Pluto can never be closer to one another than 1.5 billion miles (2.5 billion kilometers). See J. G. Williams and G. S. Benson, "Resonances in the Neptune-Pluto System," *Astronomical Journal* 76 (March 1971): 167–177.

26. James H. Applegate, Michael R. Douglas, Yekta Gürsel, Gerald J. Sussman, and Jack Wisdom, "The Outer Solar System for 200 Million Years," *Astronomical Journal* 92 (July 1986): 176–194; and Jack Wisdom, "Chaotic Dynamics in the Solar System," *Icarus* 72 (November 1987): 241–275.

27. William B. McKinnon, "On the Origin of Triton and Pluto," *Nature* 311 (September 27, 1984): 355–358.

28. William B. McKinnon and Steve Mueller, "Pluto Structure and Composition: Evidence for a Solar (vs. Planetary) Nebula Origin" (not yet published). Personal communication, April 29, 1988.

29. James W. Young, at the California Institute of Technology's Table Mountain Observatory in Wrightwood, California, has often been able to see Pluto with a 6-inch (15-centimeter) telescope.

CHAPTER 13. IS THERE A TENTH PLANET?

1. This Tombaugh quotation and others that follow may be found in Clyde W. Tombaugh and Patrick Moore, *Out of the Darkness: The Planet Pluto* (Harrisburg, Pennsylvania: Stackpole Books, 1980), pp. 130–190.

2. See "Chronology of the Search for a Tenth Planet" on page 258 of this book. For a description of earlier proposals for a tenth planet, see William Graves Hoyt, *Planets X and Pluto* (Tucson: University of Arizona Press, 1980).

3. Robert S. Harrington and Thomas C. Van Flandern, "The Satellites of Neptune and the Origin of Pluto," *Icarus* 39 (1979): 131–136.

4. Harrington, personal communication, January 25, 1988.

5. If a tenth planet is ever discovered, an appropriate name would be Persephone (or Proserpina, in Latin), after the beautiful young woman that Pluto kidnapped to be his wife. Asteroid 399, which already bears that appelation, would have to be renamed. Or maybe not. Charles T. Kowal points out that several asteroid names have been expropriated for satellites without renaming the asteroids.

6. Luis W. Alvarez, Walter Alvarez, Frank Asaro, and Helen V. Michel, "Extraterrestrial Cause for the Cretaceous-Tertiary Extinction," *Science* 208 (June 6, 1980): 1095–1108.

7. David M. Raup and J. John Sepkoski, Jr., "Periodicity of Extinctions in the Geologic Past," *Proceedings of the National Academy of Sciences USA* 81 (February 1984): 801–805.

   They found the period between extinctions to be approximately 26 million years. Another scientific team found evidence that large craters on Earth had been formed in cycles of 28 million years and that peaks in cratering activity coincided with dates of mass extinctions within the accuracy of dating methods. The cycles were in phase.

8. Daniel P. Whitmire and Albert A. Jackson IV, "Are Periodic Mass Extinctions Driven by a Distant Solar Companion?" (713–715) and Marc Davis, Piet Hut, and Richard A. Muller, "Extinctions of Species by Periodic Comet Showers" (715–717), *Nature* 308 (April 19, 1984).

   The solar companion postulated by Whitmire and Jackson was somewhat smaller and its orbit more elongated than the one postulated by Davis, Hut, and Muller.

9. See John J. Matese and Daniel P. Whitmire, "Planet X as the Source of the Periodic and Steady-State Flux of Short Period Comets," in Roman Smoluchowski, John N. Bahcall, and Mildred S. Matthews, eds., *The Galaxy and the Solar System* (Tucson: University of Arizona Press, 1986), pp. 297–309; Matese and Whitmire, "Planet X and the Origin of the Shower, and

Steady State Flux of Short-Period Comets," *Icarus* 65 (1986): 37–50; and Whitmire and Matese, "Periodic Comet Showers and Planet X," *Nature* 313 (January 3, 1985): 36–38.

10. Eugene M. Shoemaker and Ruth F. Wolfe, "Mass Extinctions, Crater Ages, and Comet Showers," in Smoluchowski, Bahcall, and Matthews, pp. 376–381.

11. John Anderson and E. Myles Standish, Jr., "Dynamic Evidence for Planet X," in Smoluchowski, Bahcall and Matthews, pp. 286–296.

12. Conley Powell, "A Mathematical Search for Planet X," to be published in the *Journal of the British Interplanetary Society* in 1988.

   Powell lists the precise numbers that emerge from his calculations but stresses that the range of uncertainty in mass, distance, period, location, and other characteristics is very great.

13. Chester explains: Without extreme cooling, the sensors warm up to the surrounding temperature and thus detect only their own heat. IRAS could then see only itself and therefore had gone blind. (Personal communication, February 11, 1988.)

14. IRAS didn't have fine resolution, so it would be possible for a tenth planet to escape detection by being too nearly in front of a star to be seen by IRAS as a separate object.

15. Brown dwarfs is an unfortunate and confusing name because these objects are not brown in color and because *dwarf* in astronomy has previously been reserved for stars or former stars, not substellar objects. These objects are distinct from red dwarfs (normal but low-mass stars) and black dwarfs (stars of moderate mass that have exhausted their ability to sustain nuclear reactions and have cooled, shrunk, and thus dimmed through white dwarfhood to black dwarfs [*not* black holes]).

16. In separate projects, Kowal discovered 81 supernovae (exploding supergiant stars) and one and a half satellites of Jupiter (Leda, moon #13, and a fourteenth moon that could not be confirmed with subsequent observations).

17. Personal communication, March 14, 1988.

18. Anderson and Standish, in Smoluchowski, Bahcall, and Matthews, pp. 286–296.

19. Clyde W. Tombaugh, "The Discovery of Pluto: Some Generally Unknown Aspects of the Story, Part II," *Mercury* 15 (July/August 1986): 98–102.

CHAPTER 14. COSMIC ARCHIVES: WHAT THE OUTER SOLAR SYSTEM CAN TELL US

1. Many of these orbits were subsequently made more circular by the tidal effects of the center of the Milky Way Galaxy and by passing stars and nebulae.

   The vast majority of primordial comets are thought to have formed in the plane of the solar system beyond Neptune and Pluto and extending outward several thousand astronomical units. This inner Oort Cloud is the reservoir from which the spherical outer Oort Cloud is replenished as passing stars and nebulae strip away the outlying comets or send them tumbling into the heart of the planetary system.

2. The Trojan asteroids at the Lagrangian positions of Jupiter may be captured comets and hence little changed (except by sunlight and the solar wind) since the solar system was formed. Another primitive body candidate is Chiron, the outermost asteroid, which may be an exceptionally large comet.

Chronologies

The Discovery of Uranus

1738, November 15	William Herschel born
1750	Caroline Herschel born
1753	Herschel joins regimental band of the Hanoverian Guards
1757, November	Herschel flees to England as French occupy Hanover
1766	Herschel appointed organist for Octagon Chapel in Bath
1772	Caroline Herschel joins her brothers William and Alexander in England
	Johann Bode reprints Titius' note about a relationship between planet distances ("Bode's Law")
1773, Spring	Herschel purchases lenses and assembles and tests refracting telescopes
Fall	Herschel is disappointed by refractors; rents a reflector; begins constructing his own reflectors
1776	*American Declaration of Independence*
1778	Herschel completes excellent 6.2-inch (7-foot focal-length) reflecting telescope; reduces his load of music students to allow more time for astronomy
1781, March 13	Herschel discovers Uranus while conducting his second all-sky survey
April 4	Nevil Maskelyne proposes that Uranus is a planet
Summer	Anders Lexell and others compute a circular orbit for Uranus, demonstrating that it is a planet
October 19	*Victory at Yorktown ends American Revolution*
November	Herschel awarded Copley Medal by the Royal Society
1782, July	Herschel given subsidy by King George III; moves to near Windsor; abandons music as profession
1783	Barnaba Oriani and others calculate first elliptical orbit for Uranus
1784	Position predictions for Uranus using orbital calculations begin to fail
1786, April	Herschel moves to Slough (near Windsor), where he lives for the rest of his life
1787, January and February	Herschel discovers the first two moons of Uranus (Titania and Oberon)
	Zach attempts and abandons effort to find "missing planet" predicted by Bode's Law
	American Constitution ratified
1788	Herschel marries widow Mary Pitt
1789	*Washington becomes first president of U.S.*
	Herschel discovers sixth and seventh moons of Saturn (Enceladus and Mimas)
1792, March 7	John Herschel born
1793	William Herschel becomes a British citizen
1800	Zach convenes "Lilienthal Detectives" and organizes search to find "missing planet" predicted by Bode's Law

1801, January 1	Guiseppe Piazzi discovers Ceres, the first asteroid
	Jefferson becomes president of U.S.
September–October	Carl Friedrich Gauss develops new method for computing orbits from a minimum of sightings
December 31	Zach recovers Ceres using position prediction by Gauss
1802, March 28	Wilhelm Olbers discovers a second asteroid: Pallas
June	Olbers suggests that asteroids are fragments of a destroyed planet
1804, September 2	Carl Ludwig Harding discovers a third asteroid: Juno
1807, March 28	Olbers discovers a fourth asteroid: Vesta
1812	*War of 1812 between U.S. and Britain begins*
1816	Herschel knighted
1821	Alexis Bouvard publishes his tables of Uranus in which he collects 17 prediscovery sightings of Uranus and then ignores them in his effort to rectify the problem with the motion of Uranus.
1822, August 25	William Herschel dies
1847	*Mexican-American War ends*
1848	Caroline Herschel dies

The Discovery of Neptune

	England	France & Germany
1781, March 13	William Herschel discovers Uranus	
1788	Uranus orbit tables fail to match position of Uranus	
1789	*Washington becomes first president of United States*	
1790		Jean B. J. Delambre publishes new Uranus tables
1811, March 11		Urbain Jean Joseph Le Verrier born
1815	*Napoleon defeated at Waterloo*	
1819, June 5	John Couch Adams born	
1821		Alexis Bouvard publishes new Uranus tables, discards prediscovery sightings for orbit calculation
1823		Friedrich Bessel criticizes Bouvard's work, especially for errors attributed to earlier astronomers
1824	*Beethoven premieres his Ninth Symphony ("Choral")*	
1826	Deviation of Uranus from Bouvard's tables becomes greater, alarming astronomers	
1829–1830	Uranus back in phase with position prediction tables	
1831	Uranus begins to fall behind its calculated place	
1831	*Charles Darwin sails on H.M.S. Beagle to South America*	
1834, November 17	Amateur astronomer T. J. Hussey proposes to George B. Airy that a planet exists beyond Uranus; Airy discourages idea that position of such a planet could be calculated	
1835	Airy named astronomer royal	Bouvard, under influence of P. A. Hansen, converted to idea that Uranus is being perturbed by an unknown planet

	England	*France & Germany*
1835	Airy proposes that gravity diverges from the inverse square law	Le Verrier becomes a chemist; publishes first scientific paper
		E. B. Valz, F. B. G. Nicolai, and Niccolò Cacciatore independently propose a planet beyond Uranus to account for perturbations in orbit of Halley's Comet
1836	Astronomers generally agree that Uranus and Halley's Comet are being disturbed by a trans-Uranian planet	
1836	*Texas wins independence from Mexico*	
1837		Le Verrier marries; becomes assistant in chemistry to Guy-Lussac at École, then assistant in astronomy
		Bouvard revises his Uranus tables with help of nephew Eugène; Airy discourages Eugène from attempt to find unknown planet
1839, September 10		Le Verrier's first astronomy publication (on variations in planetary orbits) attracts attention for its elegant mathematics
1839	*First baseball game played*	
1839, October	Adams enters Cambridge University	
1841, July 3	Adams notes intention of finding the planet that is disturbing Uranus after he graduates	
1841	*Edgar Allan Poe writes first detective story*	
1842		Bessel informs John Herschel that he intends to compute and search for unknown planet based on analysis of his pupil F. W. Flemming (but Flemming dies soon after and Bessel is ill until his death in 1846)
1843, Spring	Adams outlines plan to calculate position of a trans-Uranian planet to James Challis, director of Cambridge Observatory; Challis supportive	
1843, October	Adams produces preliminary solution, confirming existence of exterior planet	
1843	*Dickens writes* A Christmas Carol	
1844, February 13	Challis obtains new Uranus positions for Adams from Airy	
1845, Summer		François Arago urges Le Verrier to study motion of Uranus
1845, middle of September	Adams completes new solution to Uranus problem, gives summary to Challis, who urges him to send results to Airy; Adams plans to deliver them in person on way home for vacation	
1845, end of September	Adams finds that Airy is away in France	
1845, October 21	Adams tries to visit Airy again; Airy is out, then at dinner; Adams leaves summary of work	

	England	*France & Germany*
1845, November 5	Airy replies to Adams work with irrelevant question, treats work as assumption; Adams discouraged	
1845, November 10		Le Verrier presents his first paper analyzing Uranus' motion
1845	*Poe writes ''The Raven''*	
1845, June 1		Le Verrier presents his second paper on Uranus, attributing irregularities in its motion to an unknown planet
1846	*Mexican-American War begins*	
1846, June 23 or 24	Airy receives Le Verrier's second paper and notes that it agrees with Adams' prediction to within one degree	
1846, June 26	Airy responds to Le Verrier with same irrelevant question he asked Adams; doesn't mention Adams' work	
1846		Le Verrier replies to Airy by dismissing his question; offers positions if Airy will conduct search
1846, June 29	Airy proclaims to Greenwich Observatory board that a new planet will probably be discovered soon because of agreement between Adams' and Le Verrier's work	
1846, July 1	Airy declines Le Verrier's offer	
1846, July 2	Airy visits Cambridge; meets Adams by chance; no discussion of Adams' work	
1846, July 6	Airy visits former professor George Peacock, who urges him to begin a search	
1846, July 9	Airy writes Challis to request search and tell him how to do it	
1846, late July	Challis tells Adams of projected search; Adams provides updated positions; advises that planet should show disk	
1846, July 29	Challis begins search	
1846, August 12	Challis sees Neptune but fails to recognize it	
1846, August 31		Le Verrier presents third paper, providing position of disturbing planet; no search is begun
1846, September 2	Adams sends Airy his sixth solution to problem, incorporating answer to Airy's question; Airy away; Adams decides to announce his work	
1846, September 10	John Herschel predicts imminent discovery of a new planet found by its effects on Uranus	
1846, September 15	Adams tries to present findings at scientific conference; finds astronomy session has ended because of schedule change	
1846, September 18		Le Verrier writes to Johann Galle at Berlin Observatory to request a search

	England	*France & Germany*
1846, September 23		Galle receives Le Verrier's letter; receives permission to use telescope; Heinrich d'Arrest asks to help; Galle and d'Arrest find Neptune after brief search, less than one degree from predicted position
1846, September 24		Existence of new planet confirmed by its motion and disk; Galle notifies Le Verrier
1846, September 29	Challis receives Le Verrier's third paper, giving position of unknown planet; begins to look for disk (idea of Adams' he had rejected); finds object with disk but doesn't follow up at high magnification	Airy receives news of planet's discovery while traveling in Europe
1846, September 30?	Challis mentions disk he saw to W. T. Kingsley, who asks to see it; when they arrive at observatory/lodgings, Mrs. Challis insists on serving tea; sky clouds over; Challis decides not to search on following nights because of bright Moon	
1846, October 1	New planet announced in London *Times*; planet visible despite moonlight and haze; Challis examines his data and finds he saw new planet twice in first four days of observing as well as on September 29	Le Verrier notifies Galle and other astronomers falsely that French Bureau of Longitudes has named planet Neptune (actually, it is Le Verrier's initial choice)
1846, between October 1 and 5		Le Verrier changes his mind about planet's name; urges Arago to propose name Le Verrier; Arago agrees if Uranus can be changed to Herschel
1846, October 3	Herschel announces Adams' work	
1846, October 5	Challis writes Arago about his search since July 29 and his sighting of the new planet but omits mention of Adams	
1846	*Famine in Ireland as potato crop fails*	
1846, October 10	William Lassell discovers large moon of Neptune (Triton); not confirmed until July 1847	
1846, October 12	Challis notifies Airy that he saw Neptune three times but failed to recognize it	
1846, October 14	Airy writes to Le Verrier, mentioning Adams' earlier work but praising Le Verrier as discoverer because his work was more "extensive"; suggests to Challis and Adams that they write the history of the search	
1846, October 16		Le Verrier angry that his discovery is being claimed for Adams (an unknown); asks Airy why he and Challis never mentioned him and why Adams is silent
1846, October 17	Challis' account of search appears in *Athenaeum*, beginning with Adams' first solution in 1843	

	England	*France & Germany*
1846, October 19		Paris Academy of Science incensed by English claims; French press begins series of virulent attacks on England, Airy, and Challis
1846, October 22		Prominent continental astronomers decide that new planet's name should be Neptune
1846	*Sewing machine invented in U.S.*	
1846, November 13	Royal Astronomical Society holds an investigation into Neptune scandal; Airy and Challis try to justify their behavior; Adams presents summary of his work, complete with orbit he has computed on the basis of Challis' sightings of Neptune; Airy and Challis receive heavy criticism	
1846, November 30	Royal Society awards Le Verrier its Copley Medal although they know of Adams' work	
1846, late	Royal Astronomical Society decides to award no prize for 1846	
1847, June	Adams and Le Verrier meet for first time, arranged by Herschel; they form a lifelong friendship	
1848	Royal Society awards Adams the Copley Medal; Royal Astronomical Society gives testimonials to 12 scientists, including Adams, Le Verrier, and Airy, but none to Galle, d'Arrest, or Challis	

The Discovery of Pluto

1855, March 13	Percival Lowell born in Boston
1861	*American Civil War begins*
1876	Lowell graduates from Harvard with bachelor's degree in mathematics; following a year's travel in Europe, he is employed by his grandfather
1883	Lowell, now wealthy, leaves his grandfather's company; begins extended visits to the Far East as a travel writer
1883	*"Buffalo Bill" Cody organizes his "Wild West Show"*
1893	Lowell returns to Boston to establish an observatory to study Mars and other planets
1893	*Dvořák writes his Ninth Symphony ("From the New World")*
1894, June 1	Lowell Observatory opens in Flagstaff, Arizona
1905, February	Lowell begins sporadic and secret telescopic and mathematical search for a trans-Neptunian planet
1905	*Einstein publishes his Special Theory of Relativity*
1906, February 4	Clyde William Tombaugh born on farm near Streator, Illinois
1906	*Great San Francisco earthquake*
1907, September 6	The photographic survey component of Lowell's first search for a ninth planet ends unsuccessfully

1908, June 10	Lowell, age 53, marries neighbor Constance Savage Keith
1908, November 11	William H. Pickering, using a graphical plot, announces the first of his many predictions of a trans-Neptunian planet (Planet O); two immediate searches for this planet find nothing; Lowell pours new energy into his mathematical analysis, adopting the graphical method
1909, May	Lowell calculates position of a ninth planet but does not publish his work or search this location; first search for his Planet X ends
1910, July	Lowell begins second quest for ninth planet with extensive mathematical analysis to guide telescopic search
1911, March 13	Second telescopic search for Planet X begins at Lowell Observatory based on Lowell's mathematical predictions
1911	Lowell buys blink microscope to facilitate search on photographic plates for ninth planet
1912, October	Lowell collapses from nervous exhaustion due to mathematical work on trans-Neptunian planet; resumes part-time work two months later
1914	*World War I begins*
1915, January 13	Lowell publishes his mathematical and observational efforts to find a ninth planet; report generates little scientific or public interest
1915	Continuing Lowell Observatory search for Planet X records Pluto very faintly on photographs made on March 19 and April 7; planet goes undetected
1916, July 2	Lowell's second search for trans-Neptunian planet ends without success
1916	*Einstein publishes his General Theory of Relativity*
1916, November 12	Percival Lowell, age 61, dies of a stroke at the Lowell Observatory
1918	*World War I ends*
1919	Pickering publishes a new prediction for his Planet O; Milton Humason performs two photographic searches at the Mount Wilson Observatory and Pluto is barely recorded on four plates but goes unnoticed
1925	Tombaugh completes high school in Kansas; farms during summers and builds telescopes during winters
1927	A. Lawrence Lowell contributes funds for 13-inch refractor to resume search for Lowell's Planet X
1928	Pickering publishes a new prediction for his Planet O; a Harvard Observatory search fails to find it
1929, January 15	Tombaugh arrives at Lowell Observatory to take up duties as an assistant observer on a trial basis
1929, April 6	Third search for ninth planet begins at the Lowell Observatory with Clyde W. Tombaugh, age 23, at the telescope
1929, Summer	Tombaugh assumes responsibility for blink comparator use in the Planet X project because analysis of his photographic search plates is far behind
1929, October 28	*U.S. stock market collapses*
1930, January 21, 23, and 29	Trans-Neptunian planet photographed by Tombaugh near Delta Geminorum
1930, February 18	Tombaugh, age 24, finds ninth planet while examining January 23 and 29 plates with blink microscope; Slipher delays announcement to confirm discovery, to begin calculation of planet's orbit, and to avoid immediate upstaging by larger observatories
1930, March 13	Lowell Observatory announces discovery of trans-Neptunian planet on 75th anniversary of Lowell's birth and 149th anniversary of the discovery of Uranus
1930, April 12	Lowell Observatory announces a provisional orbit for ninth planet and finally releases additional positions for Pluto
1930, May 1	Lowell Observatory proposes the name Pluto for the new planet, as suggested by Venetia Burney, an 11-year-old English schoolgirl
1930, May	Tombaugh resumes photographic and blink comparator search for additional trans-Neptunian planets (search continues 13 years)

1930	*Grant Wood paints American Gothic*
1934	Issei Yamamoto (Japan) suggests that Pluto is an escaped satellite of Neptune
1936	Raymond A. Lyttleton (England) suggests that Pluto was a satellite of Neptune ejected from the system by a near collision with Triton that caused Triton to assume a retrograde orbit
1939	*World War II begins in Europe*
1943	Tombaugh leaves Lowell Observatory staff for wartime duty as a navigation instructor and then a variety of other science and teaching positions
1950	Measurements by Gerard P. Kuiper with the 200-inch telescope at Palomar set upper limit for Pluto's diameter at 3,600 miles (5,900 kilometers)
1950	*Korean War begins*
1953	M. F. Walker and R. Hardie measure Pluto's rotation at 6.39 days
1955	Tombaugh joins faculty at New Mexico State University
1969	*First man on the Moon*
1971	J. G. Williams and G. S. Benson demonstrate that Pluto and Neptune cannot collide
1976	Infrared observations of Pluto by Dale P. Cruikshank, Carl B. Pilcher, and David Morrison indicate that its surface is covered by reflective methane frost, suggesting that Pluto is smaller than Earth's Moon
1977	*Voyager 2 launched on Grand Tour of giant planets*
1978, June	James W. Christy discovers a satellite of Pluto on photographs taken at the U.S. Naval Observatory's Flagstaff station on April 13 and 20 and May 12, 1978; moon quickly named Charon; discovery reconfirms Pluto's small size and provides proof that Pluto could not have caused perturbations on Uranus that spurred search by Lowell and Pickering
1979, January 21	Pluto's orbit brings it closer to the Sun than Neptune; will remain closer until 1999
1985	Pluto and Charon begin to eclipse one another as seen from Earth
1986, January 24	*Voyager 2 passes Uranus*
1989, August 24	*Voyager 2 passes Neptune*
1989, September 12	Pluto reaches perihelion (its closest approach to the Sun)
1999, March 14	Pluto's orbit returns it to a distance farther from the Sun than Neptune for the next 228 years

Modern Discoveries About the Outer Planets

1948	Discovery of Miranda, fifth moon of Uranus, by Gerard P. Kuiper
1949	Discovery of Nereid, second moon of Neptune, by Gerard P. Kuiper
1950	Existence of vast spherical cloud of comets surrounding the solar system (the Oort Cloud) is proposed by Jan H. Oort (a similar proposal was made by Ernst Öpik in 1932)
1950	Comets as accretions of ice and dust-sized rock particles ("dirty snowballs") proposed by Fred L. Whipple
1950	*Korean War begins*
1965	Gary A. Flandro discovers method of reaching the outer planets through successive gravitational assists—the Grand Tour concept
1969	*First man on Moon*

1971	NASA's plan for Grand Tour mission to Jupiter, Saturn, Uranus, and Neptune is canceled as NASA budget shrinks and Space Shuttle requires more funds
1972, March 2	NASA's *Pioneer 10* launched toward Jupiter; will use the gravity of Jupiter to accelerate out of the solar system
1972	NASA funds two *Mariner Jupiter-Saturn* (later *Voyager*) gravity-assist missions to replace Grand Tour project
1973, April 5	NASA's *Pioneer 11,* the first mission to use a gravity assist to reach an outer planet, launched toward Jupiter; uses Jupiter's gravity to accelerate on to Saturn
1973, November 3	NASA's *Mariner 10* is launched to Venus and Mercury; becomes the first mission to employ gravity assist by using the gravity of Venus (February 5, 1974) to slow spacecraft so that it falls sunward for three encounters with Mercury (beginning March 29, 1974)
1973, December 3	*Pioneer 10* becomes the first spacecraft to reach Jupiter
1973	NASA begins preparation for a third *Voyager* gravity-assist mission to Jupiter and Uranus
1974	*Worldwide inflation*
1974, December 2	*Pioneer 11* passes Jupiter and acquires gravitational boost toward Saturn
1975	*Voyager 3* mission to Jupiter and Uranus canceled
1975	Methane and nitrogen discovered on Triton
1976	Methane ice detected on surface of Pluto
1976	A Chorus Line *opens on Broadway*
1977	Rings of Uranus discovered by James Elliot and colleagues
1977	*Voyager 2* (August 20) and *Voyager 1* (September 5; on a shorter flight path) are launched toward Jupiter for gravity assist to Saturn; *Voyager 2*'s launch is timed to allow a Grand Tour mission on to Uranus and Neptune if *Voyager 1* succeeds at Jupiter and Saturn
1978, June 22	Charon, moon of Pluto, discovered by James W. Christy
1979, January 21	Pluto crosses inside orbit of Neptune, passing 8.9 astronomical units above Neptune's path, leaving Neptune as the outermost planet for 20 years
1979, March 5	*Voyager 1* passes Jupiter; discovers its ring system
1979, July 9	*Voyager 2* passes Jupiter
1979, September 1	*Pioneer 11* becomes the first spacecraft to reach Saturn
1980, November 12	*Voyager 1* successfully passes Saturn; *Voyager 2* able to undertake Grand Tour mission on to Uranus and Neptune
1981, April 12	*First Space Shuttle launch*
1981, August 25	*Voyager 2* passes Saturn and acquires gravitational boost on to Uranus, continuing hopes for four-planet Grand Tour
1984	Partial ring system of Neptune detected
1985	Pluto and Charon begin to eclipse one another as seen from Earth
1986, January 24	*Voyager 2* becomes the first spacecraft to reach Uranus; acquires gravitational boost on to Neptune
1986, January 28	*Space Shuttle* Challenger *explodes shortly after launch, killing crew of seven*
1987	Water ice discovered on Charon
1989, August 24	*Voyager 2* becomes the first spacecraft to reach Neptune
1989, September 12	Pluto reaches perihelion (closest approach to the Sun): 29.6 astronomical units (2.75 billion miles; 4.43 billion kilometers)
1990, October	Pluto and Charon no longer eclipse one another as seen from Earth; next eclipse season will begin about 2109
1999, March 14	Pluto crosses 6.1 astronomical units above the orbit of Neptune outbound to resume its usual role as the outermost planet

The Search for a Tenth Planet

1929 to 1943	Clyde Tombaugh surveys the entire sky as seen from the Lowell Observatory in Arizona with a 13-inch (33-centimeter) refractor and a blink comparator in search of trans-Neptunian objects (Pluto, the ninth planet, is discovered in 1930)
1942	Robert S. Richardson (U.S.) proposes a tenth planet at 36 astronomical units to explain the delay of Halley's Comet in reaching perihelion
1946	M. E. Sevin (France) predicts a tenth planet at 78 astronomical units
1950	K. Schuette (West Germany) proposes a tenth planet at 77 astronomical units based on comet orbits
1954	H. H. Kritzinger (West Germany) refines Schuette's work and predicts a tenth planet at 65 astronomical units
1957	Kritzinger redoes his work and predicts a tenth planet at 75 astronomical units
1959	Kritzinger reanalyzes his work and proposes a tenth planet at 77 astronomical units (Schuette's original distance)
1972	Joseph L. Brady (U.S.) predicts a tenth planet at 60 astronomical units in a retrograde orbit with a mass 285 times that of Earth
1975	Gleb Chebotarev (Soviet Union) predicts two undiscovered planets at 54 and 100 astronomical units based on an analysis of comet orbits
1977 to 1984	Charles T. Kowal conducts a systematic search 15 degrees on either side of the ecliptic for undiscovered bodies in the solar system using Palomar Observatory's 48-inch (122-centimeter) Schmidt Telescope and a blink comparator
1977	Kowal discovers Chiron, about 150 miles (250 kilometers) in diameter, the outermost asteroid known, with perihelion inside the orbit of Saturn (8.5 astronomical units) and aphelion inside the orbit of Uranus (18.9 astronomical units). Initial press coverage treats it as a tenth planet
1979	Robert S. Harrington and Thomas C. Van Flandern (U.S.) propose a tenth planet ("Humphrey") in a highly inclined and eccentric orbit at a distance of 50 to 100 astronomical units, a period of about 800 years, and a mass of 2 to 5 Earths to explain the remaining irregularities in the motion of Uranus and Neptune, the unusual orbits of Neptune's moons, and the anomalous orbit of Pluto. Harrington commences a search for the suspected planet
1985	Daniel P. Whitmire and John J. Matese (U.S.) propose a tenth planet to explain periodic mass extinctions on Earth due to comet showers; planet orbits at about 80 astronomical units with substantial ellipticity and inclination and a mass of 1 to 5 Earths
1987	John D. Anderson (U.S.) analyzes *Pioneer 10* and *11* spacecraft trajectories for evidence of perturbations by a tenth planet; finds none but concludes that a tenth planet must exist in a highly inclined orbit to account for the discrepancies in the motion of Uranus and Neptune
1987	Conley Powell (U.S.) predicts the position of a tenth planet in a nearly circular, low-inclination orbit on the basis of a mathematical analysis of the Uranus residuals. Edward Bowell at the Lowell Observatory directs a search of Tombaugh's plates for the object, but nothing is found
1988	Thomas J. Chester and Michael Melnyk (U.S.) complete a two-year review of images from the Infrared Astronomical Satellite (IRAS) covering one-tenth of the sky in search of a tenth planet; they eventually plan to search one-half of the sky; project is part of a search for substellar objects in the solar neighborhood

Voyager 1 and 2

Spacecraft Data

Spacecraft and planet
encounter data supplied by
Rex W. Ridenoure
Voyager Mission Planning Office
Jet Propulsion Laboratory

Interstellar mission
data supplied by
Robert J. Cesarone
Voyager Navigation Team
Jet Propulsion Laboratory

Voyager 1 and 2

Total Weight	1,817 pounds (824 kilograms)
Scientific Payload	232 pounds (106 kilograms)
Height of Spacecraft	12 feet (3.7 meters)
Width of Receiver/Transmitter Dish	12 feet (3.7 meters)
Maximum Spacecraft Width with Booms Extended	57 feet (17.3 meters)
Launch Vehicle	Titan-Centaur

	Voyager 1	**Voyager 2**
Launch	September 5, 1977	August 20, 1977
(*Voyager 2* lifted off before *Voyager 1* but traveled a longer, slower path)		
Jupiter Encounter		
Closest approach	March 5, 1979	July 9, 1979
Distance from center of planet	216,790 miles (348,890 kilometers)	448,425 miles (721,670 kilometers)
Distance from cloud tops	172,425 miles (277,490 kilometers)	404,060 miles (650,270 kilometers)
Distance from Earth	422 million miles (679 million kilometers)	579 million miles (932 million kilometers)
Time for radio signal to reach Earth	37.7 minutes	51.8 minutes

	Voyager 1	Voyager 2
Saturn Encounter		
Closest approach	November 12, 1980	August 25, 1981
Distance from center of planet	114,440 miles (184,175 kilometers)	100,100 miles (161,095 kilometers)
Distance from cloud tops	77,160 miles (124,175 kilometers)	62,815 miles (101,095 kilometers)
Distance from Earth	948 million miles (1.52 billion kilometers)	968 million miles (1.56 billion kilometers)
Time for radio signal to reach Earth	1 hour 25 minutes	1 hour 27 minutes
Uranus Encounter		
Closest approach	-none-	January 24, 1986
Distance from center of planet		66,500 miles (107,100 kilometers)
Distance from cloud tops		50,700 miles (81,600 kilometers)
Distance from Earth		1.84 billion miles (2.97 billion kilometers)
Time for radio signal to reach Earth		2 hours 45 minutes
Neptune Encounter		
Closest approach	-none-	August 24, 1989
Distance from center of planet		18,133 miles (29,183 kilometers)
Distance from cloud tops		2,700 miles (4,400 kilometers)
Distance from Earth		2.75 billion miles (4.42 billion kilometers)
Time for radio signal to reach Earth		4 hours 6 minutes
Interstellar Mission		
Direction headed after last encounter	Ophiuchus, the Serpent Bearer	Pavo, the Peacock
Asymptotic velocity	37,000 miles per hour (16.6 kilometers per second)	33,000 miles per hour (14.8 kilometers per second)
First stellar encounter	AC+79 3888 in 40,272 years	Ross 248 in 40,176 years
Encounter distance	1.6 light-years	1.7 light-years
Present location of encounter star	Camelopardalis, the Giraffe	Andromeda, the Princess

Spacecraft Instrumentation

Science Instruments	Data Sought
Cameras (2)	Color and black-and-white (high-resolution) images of planets, moons, and rings; search for new satellites and rings
Photopolarimeter	Size and composition of particles in rings and atmospheres
Infrared Interferometer Spectrometer and Radiometer	Composition and temperature of atmospheres; composition and temperature of moon surfaces
Ultraviolet Spectrometer	Atmospheric composition; size and composition of rings
Magnetometers (3)	Charged particles (magnetosphere) around planet
Cosmic Ray Detector	Charged particles from other stars
Plasma Detector	Charged particles around planet
Plasma Wave Detector	Charged particles around planet
Low-Energy Charged Particle Detector	Charged particles (magnetosphere) around planet; tiny ring particles
Planetary Radio Astronomy	Tiny ring particles; magnetic fields
Radio Transmissions	Composition and structure of atmosphere and rings

Solar System Statistics

Orbital Data for Planets

Planet	Mean Distance from Sun			Period of Revolution		Eccentricity	Inclination (degrees)
	Astronomical Units	Miles (in millions)	Kilometers (in millions)	Sidereal	Synodic (days)		
Mercury	0.39	36.0	57.9	87.97 days	116	0.206	7.0
Venus	0.72	67.2	108.2	224.70 days	584	0.007	3.4
Earth	1.00	93.0	149.6	365.26 days	—	0.017	0.0
Mars	1.52	141.5	227.9	686.98 days	780	0.093	1.8
Jupiter	5.20	483.3	778.3	11.86 years	399	0.048	1.3
Saturn	9.54	886.2	1427.0	29.46 years	378	0.056	2.5
Uranus	19.18	1782.0	2869.6	84.01 years	370	0.047	0.8
Neptune	30.06	2792.4	4496.6	164.79 years	367	0.009	1.8
Pluto	39.44	3663.8	5899.9	247.69 years	367	0.250	17.2

Physical Data for Sun, Moon, and Planets

Object	Equatorial Diameter Miles	Equatorial Diameter Kilometers	Mass (Earth = 1)	Mean Density (water = 1)	Surface Gravity (Earth = 1)
Sun	865,000	1,392,000	332,946.0	1.41	27.9
Mercury	3,031	4,878	0.055	5.43	0.38
Venus	7,521	12,104	0.815	5.24	0.91
Earth	7,926	12,756	1.000	5.52	1.00
Mars	4,217	6,787	0.107	3.94	0.38
Jupiter	88,700	142,800	317.833	1.33	2.54
Saturn	74,600	120,000	95.159	0.70	1.08
Uranus	31,800	51,200	14.500	1.30	0.91
Neptune	30,800	49,600	17.204	1.76	1.19
Pluto	1,457	2,345	0.002	2.00	0.05
Earth's Moon	2,160	3,476	0.012	3.34	0.17

Object	Escape Speed Miles per second	Escape Speed Kilometers per second	Rotation Period (days)	Oblateness	Inclination of Equator to Orbit (degrees)	Albedo
Sun	383.7	617.5	25[1]	0	—	—
Mercury	2.7	4.3	58.65	0	0.0	0.11
Venus	6.5	10.4	243.02	0	177.3 (retrograde)	0.65
Earth	7.0	11.2	1.00	1/298	23.4	0.37
Mars	3.1	5.0	1.03	1/193	25.2	0.15
Jupiter	37.0	59.6	0.41[2]	1/15	3.1	0.52
Saturn	22.1	35.6	0.44	1/9	26.7	0.47
Uranus	13.2	21.3	0.72	1/45	97.9 (retrograde)	0.51
Neptune	14.8	23.8	0.71	1/40	29.6	0.41
Pluto	0.7	1.2	6.39	(0)	94.0 (retrograde)	0.54
Earth's Moon	1.5	2.4	27.32	0	6.7	0.12

[1]Varies from 25–35 depending on latitude.
[2]At equator; slower toward poles.

Planet	Maximum Surface Temperature	Atmosphere (major constituents)	Atmospheric Pressure at Surface (Earth = 1)	Number of Confirmed Moons (1988)
Mercury	800°F (430°C)	negligible	negligible	0
Venus	900°F (480°C)	CO_2, N_2 clouds of H_2SO_4 & H_2O	90	0
Earth	136°F (58°C)	N_2, O_2, A clouds of H_2O	1	1
Mars	80°F (28°C)	CO_2, N_2, A clouds of H_2O & CO_2	0.007	2
Jupiter	−260°F (−160°C) cloud tops	H, He, CH_4, NH_3, H_2O clouds of NH_3, NH_4SH (ammonium hydrosulfide), & H_2O		16
Saturn	−290°F (−180°C) cloud tops	H, He, CH_4 clouds of NH_3, NH_4SH (ammonium hydrosulfide), & H_2O		17
Uranus	−357°F (−216°C) cloud tops	H, He, CH_4 clouds of CH_4		15
Neptune	−357°F (−216°C) cloud tops	H, He, CH_4 clouds of CH_4		2
Pluto	−370°F (−223°C) cloud tops	CH_4 (very tenuous)		1

Orbital Data for Moons

Name	Mean Distance from Planet[1]		Period of Revolution (days)	Eccentricity	Inclination of Orbit (degrees)
	Miles	Kilometers			
Satellite of Earth					
Moon	238,900	384,500	27.32 (sidereal)	0.055	18–29
			29.53 (synodic)		
Satellites of Mars					
Phobos	5,840	9,400	0.319	0.015	1.1
Deimos	14,600	23,500	1,263	0.001	1.8 (varies)
Satellites of Jupiter					
Metis (XVI)[2]	79,500	128,000	0.294	0	0
Adrastea (XV)	80,000	129,000	0.297	0	0
Amalthea (V)	112,000	180,000	0.498	0.003	0.4
Thebe (XIV)	138,000	222,000	0.674	0.013	0
Io (I)	262,000	422,000	1.769	0.004	0
Europa (II)	417,000	671,000	3.551	0.010	0.5
Ganymede (III)	665,000	1,070,000	7.155	0.001	0.2
Callisto (IV)	1,171,000	1,885,000	16.689	0.007	0.2
Leda (XIII)	6,698,000	11,110,000	240	0.147	27
Himalia (VI)	7,127,000	11,470,000	251	0.158	28
Lysithea (X)	7,276,000	11,710,000	260	0.107	29
Elara (VII)	7,295,000	11,740,000	260	0.207	28
Ananke (XII)	13,173,000	21,200,000	631	0.17	147 (retrograde)
Carme (XI)	13,888,000	22,350,000	692	0.21	164 (retrograde)
Pasiphae (VIII)	14,497,000	23,330,000	735	0.38	148 (retrograde)
Sinope (IX)	14,521,000	23,370,000	758	0.28	153 (retrograde)
Satellites of Saturn					
Atlas (XV)	85,100	137,000	0.601	0.002	0.3
Prometheus (XVI)	86,400	139,000	0.613	0.002	0.0
Pandora (XVII)	88,200	142,000	0.628	0.004	0.1
Janus (X)	93,800	151,000	0.695	0.009	0.3
(coorbital satellite with Epimetheus)					
Epimetheus (XI)	93,800	151,000	0.695	0.007	0.1
(coorbital satellite with Janus)					
Mimas (I)	116,000	187,000	0.942	0.020	1.5
Enceladus (II)	148,000	238,000	1.370	0.004	0.02
Tethys (III)	183,000	295,000	1.888	0.000	1.1
Telesto (XIII)	183,000	295,000	1.888	~0	~0
(librates about the trailing Lagrangian Point (L5) of Tethys' orbit)					

Name	Mean Distance from Planet[1]		Period of Revolution (days)	Eccentricity	Inclination of Orbit (degrees)
	Miles	Kilometers			
Calypso (XIV)	183,000	295,000	1.888	~0	~0
(librates about the leading Lagrangian Point (L4) of Tethys' orbit)					
Dione (IV)	235,000	378,000	2.737	0.002	0.02
Helene (XII)	235,000	378,000	2.737	0.005	0.2
(librates about the leading Lagrangian Point (L4) of Dione's orbit)					
Rhea (V)	327,000	526,000	4.517	0.001	0.4
Titan (VI)	759,000	1,221,000	15.945	0.029	0.3
Hyperion (VII)	920,000	1,481,000	21.276	0.104	0.4
Iapetus (VIII)	2,213,000	3,561,000	79.331	0.028	14.7
Phoebe (IX)	8,053,000	12,960,000	550.46	0.163	150 (retrograde)

Satellites of Uranus

Name	Miles	Kilometers	Period of Revolution (days)	Eccentricity	Inclination of Orbit (degrees)
1986U7[3]	30,900	49,700	0.333	~0	~0
(inner Epsilon Ring shepherd)					
1986U8	33,400	53,800	0.375	~0	~0
(outer Epsilon Ring shepherd)					
1986U9	36,800	59,200	0.433	~0	~0
1986U3	38,400	61,800	0.463	~0	~0
1986U6	39,000	62,700	0.475	~0	~0
1986U2	40,100	64,600	0.492	~0	~0
1986U1	41,100	66,100	0.513	~0	~0
1986U4	43,400	69,900	0.558	~0	~0
1986U5	46,800	75,300	0.621	~0	~0
1985U1 (Puck)	53,400	86,000	0.763	~0	~0
Miranda (V)	80,700	129,900	1.413	0.017	3.4
Ariel (I)	118,600	190,900	2.251	0.003	0
Umbriel (II)	165,300	266,000	4.146	0.004	0
Titania (III)	271,100	436,300	8.704	0.002	0
Oberon (IV)	362,500	583,400	13.463	0.001	0

Satellites of Neptune

Name	Miles	Kilometers	Period of Revolution (days)	Eccentricity	Inclination of Orbit (degrees)
Triton (I)	220,000	354,000	5.877	0.000	160 (retrograde)
Nereid (II)	3,500,000	5,600,000	365.21	0.75	27.6

Satellite of Pluto

Name	Miles	Kilometers	Period of Revolution (days)	Eccentricity	Inclination of Orbit (degrees)
Charon	12,400	20,000	6.387	~0	~0

[1]Measured from center of planet to center of moon.

[2]An old attempt to number each planet's satellites in order from closest to farthest for ease of reference. Thus Io is Jupiter I. The intent of this system has been disrupted by more recent satellite discoveries.

[3]That is, the seventh satellite of Uranus to be discovered in 1986.

Physical Data for Moons

	Diameter		Mass (Earth's Moon = 1)	Density (water = 1)	Albedo	Discovery
	Miles	Kilometers				
Satellite of Earth						
Moon	2,160	3,476	[734.9×10^{20} kg]	3.34	0.11	
Satellites of Mars						
Phobos	13	21	1.8×10^{-7}	~2	0.07	A. Hall, 1877
Deimos	7	12	2.4×10^{-8}	~2	0.07	A. Hall, 1877
Satellites of Jupiter						
Metis (XVI)[1]	(25)	(40)			(0.05)	S. Synnott, 1979
Adrastea (XV)	(15)	(25)			(0.05)	Jewitt, Danielson, Synnott, 1979
Amalthea (V)	106	170			0.05	E. Barnard, 1892
Thebe (XIV)	(60)	(100)			(0.05)	S. Synnott, 1979
Io (I)	2,260	3,630	1.21	3.55	0.6	Galileo, 1610
Europa (II)	1,950	3,140	0.66	3.04	0.6	Galileo, 1610
Ganymede (III)	3,270	5,260	2.03	1.93	0.4	Galileo, 1610
Callisto (IV)	2,980	4,800	1.46	1.83	0.2	Galileo, 1610
Leda (XIII)	(10)	(15)				C. Kowal, 1974
Himalia (VI)	115	185			0.03	C. Perrine, 1904
Lysithea (X)	(20)	(35)				S. Nicholson, 1938
Elara (VII)	45	75			0.03	C. Perrine, 1905
Ananke (XII)	(20)	(30)				S. Nicholson, 1951
Carme (XI)	(25)	(40)				S. Nicholson, 1938
Pasiphae (VIII)	(30)	(50)				P. Melotte, 1908
Sinope (IX)	(20)	(35)				S. Nicholson, 1914
Satellites of Saturn						
Atlas (XV)	20	30			0.4	R. Terrile, 1980
Prometheus (XVI)	60	100			0.6	S. Collins, D. Carlson, 1980
Pandora (XVII)	55	90			0.5	S. Collins, D. Carlson, 1980
Janus (X) (coorbital satellite with Epimetheus)	120	190			0.3	A. Dollfus, 1966
Epimetheus (XI) (coorbital satellite with Janus)	75	120			0.5	J. Fountain, S. Larson, 1966
Mimas (I)	240	390	5.2×10^{-4}	1.2	0.8	W. Herschel, 1789
Enceladus (II)	310	500	~0.001	1.1	1.0	W. Herschel, 1789
Tethys (III)	660	1,060	0.010	1.2	0.8	G. Cassini, 1684
Telesto (XIII) (librates about the trailing Lagrangian Point (L5) of Tethys' orbit)	15	25			0.7	Smith, Larson, Reitsma, 1980

	Diameter		Mass (Earth's Moon = 1)	Density (water = 1)	Albedo	Discovery
	Miles	Kilometers				
Calypso (XIV)	15	25			1.0	Pascu, Seidelmann, Baum, Currie, 1980
(librates about the leading Lagrangian Point (L4) of Tethys' orbit)						
Dione (IV)	695	1,120	0.014	1.4	0.6	G. Cassini, 1684
Helene (XII)	20	30			0.6	P. Laques, J. Lecacheux, 1980
(librates about leading Lagrangian Point (L4) of Dione's orbit)						
Rhea (V)	950	1,530	0.034	1.3	0.6	G. Cassini, 1672
Titan (VI)	3,200	5,150	1.83	1.88	0.2	C. Huygens, 1655
Hyperion (VII)	160	255			0.3	W. Bond, G. Bond, W. Lassell, 1848
Iapetus (VIII)	905	1,460	0.026	1.2	0.08	G. Cassini, 1671
Phoebe (IX)	135	220			0.05	W. Pickering, 1898

Satellites of Uranus

	Diameter		Mass (Earth's Moon = 1)	Density (water = 1)	Albedo	Discovery
	Miles	Kilometers				
1986U7[2]	(25)	(40)			<0.1	*Voyager 2, 1986*
(inner Epsilon Ring shepherd)						
1986U8	(30)	(50)			<0.1	*Voyager 2, 1986*
(outer Epsilon Ring shepherd)						
1986U9	(30)	(50)			<0.1	*Voyager 2, 1986*
1986U3	(35)	(60)			<0.1	*Voyager 2, 1986*
1986U6	(35)	(60)			<0.1	*Voyager 2, 1986*
1986U2	(50)	(80)			<0.1	*Voyager 2, 1986*
1986U1	(50)	(80)			<0.1	*Voyager 2, 1986*
1986U4	(35)	(60)			<0.1	*Voyager 2, 1986*
1986U5	(35)	(60)			<0.1	*Voyager 2, 1986*
1985U1 (Puck)	105	170			<0.1	*Voyager 2, 1985*
Miranda (V)	300	485	0.001	1.3	0.34	G. Kuiper, 1948
Ariel (I)	720	1,160	0.018	1.7	0.40	W. Lassell, 1851
Umbriel (II)	740	1,190	0.017	1.4	0.19	W. Lassell, 1851
Titania (III)	1,000	1,610	0.047	1.6	0.28	W. Herschel, 1787
Oberon (IV)	965	1,550	0.040	1.5	0.24	W. Herschel, 1787

Satellites of Neptune

	Diameter		Mass (Earth's Moon = 1)	Density (water = 1)	Albedo	Discovery
	Miles	Kilometers				
Triton (I)	(2,170)	(3,500)	~ 1.8		(0.4)	W. Lassell, 1846
Nereid (II)	(190)	(300)				G. Kuiper, 1949

Satellite of Pluto

	Diameter		Mass (Earth's Moon = 1)	Density (water = 1)	Albedo	Discovery
	Miles	Kilometers				
Charon	775	1,250	0.03	~ 2	0.34	J. Christy, 1978

[1]An old attempt to number each planet's satellites in order from closest to farthest for ease of reference. This Io is Jupiter I. The intent of this system has been disrupted by more recent satellite discoveries.

[2]That is, the seventh satellite of Uranus to be discovered in 1986.

Selected
Bibliography

Adams, John Couch. *The Scientific Papers*, edited by William Grylls Adams, with a biographical notice by J. W. L. Glaisher. 2 vols. (Cambridge: Cambridge University Press, 1896, 1900).

Airy, George B. "Account of Some Circumstances Historically Connected with the Discovery of the Planet Exterior to Uranus," *Monthly Notices* of the Royal Astronomical Society 7 (1846).

Airy, George Biddell. *Autobiography*, edited by Wilfrid Airy (Cambridge: At the University Press, 1896).

Alexander, A. F. O'D. *The Planet Uranus: A History of Observation, Theory, and Discovery* (New York: American Elsevier, 1965).

Alvarez, Luis W., Walter Alvarez, Frank Asaro, and Helen V. Michel. "Extraterrestrial Cause for the Cretaceous-Tertiary Extinction," *Science* 208 (June 6, 1980): 1095–1108.

Beatty, J. Kelly, Brian O'Leary, and Andrew Chaikin, eds. *The New Solar System*, 2d ed. (Cambridge: Cambridge University Press and Cambridge, Mass.: Sky Publishing, 1982).

Binzel, Richard P., et al. "The Detection of Eclipses in the Pluto-Charon System," *Science* 228 (June 7, 1985): 1193–1195.

Brown, Robert Hamilton. "Exploring the Uranian Satellites," *Planetary Report* 6 (November/December 1986): 4–7, 18.

Cannon, Annie J. "William Henry Pickering: 1858–1938," *Science* 87 (February 25, 1938): 179–180.

Cesarone, Robert J., Andrey B. Sergeyevsky, and Stuart J. Kerridge. "Prospects for the *Voyager* Extra-Planetary and Interstellar Mission," *Journal of the British Interplanetary Society* 37 (March 1984): 99–116.

Challis, James. "The Search for the Planet Neptune by Professor Challis," *Astronomische Nachricten* 583 (1846): cols. 101–106.

Challis, James. "Special Report of Proceedings in the Observatory Relative to the New Planet," *Monthly Notices* of the Royal Astronomical Society 7 (1846).

Christy, James W., and Robert S. Harrington. "The Satellite of Pluto," *Astronomical Journal* 83 (August 1978): 1005–1008.

Clerke, Agnes M. *A Popular History of Astronomy during the Nineteenth Century*, 4th ed. (London: Adam and Charles Black, 1902).

Davis, Joel. *Flyby: The Interplanetary Odyssey of Voyager 2* (New York: Atheneum, 1987).

Davis, Marc, Piet Hut, and Richard A. Muller. "Extinctions of Species by Periodic Comet Showers," *Nature* 308 (April 19, 1984): 715–717.

DeVorkin, David H. *The History of Modern Astronomy and Astrophysics: A Selected, Annotated Bibliography* (New York: Garland Publishing, 1982).

Drake, Stillman, and Charles T. Kowal. "Galileo's Sighting of Neptune," *Scientific American* 243 (December 1980): 74–81.

Elliot, James L., E. Dunham, and D. Mink. "The Rings of Uranus," *Nature* 267 (May 26, 1977): 328–330.

Elliot, James, and Richard Kerr. *Rings: Discoveries from Galileo to Voyager* (Cambridge: MIT Press, 1984).

Goldreich, Peter, and Scott Tremaine. "Towards a Theory for the Uranian Rings," *Nature* 277 (January 11, 1979): 97–99.

Goldreich, Peter, Scott Tremaine, and Nicole Borderies. "Towards a Theory for Neptune's Arc Rings," *Astronomical Journal* 92 (August 1986): 490–494.

Greenberg, Richard, and André Brahic, eds. *Planetary Rings* (Tucson: University of Arizona Press, 1984).

Grosser, Martin. *The Discovery of Neptune* (Cambridge: Harvard University Press, 1962).

Grosser, Martin. "The Search for a Planet beyond Neptune," *Isis* 55 (August 1964): 163–183.

Harrington, Robert S., and Betty J. Harrington. "The Discovery of Pluto's Moon," *Mercury* 8 (January/February 1979): 1–3, 6, 17.

Harrington, Robert S., and Betty J. Harrington, "Pluto: Still an Enigma After 50 Years," *Sky & Telescope* 59 (June 1980): 452–454.

Harrington, Robert S., and Thomas C. Van Flandern. "The Satellites of Neptune and the Origin of Pluto," *Icarus* 39 (1979): 131–136.

Herschel, John. "Le Verrier's Planet," *Athenaeum* (October 3, 1846): 1019.

Herschel, John F. W. *Outlines of Astronomy* (Philadelphia: Lea & Blanchard, 1849).

Herschel, Mary Cornwallis [Mrs. John]. *Memoir and Correspondence of Caroline Herschel* (New York: D. Appleton, 1878).

Herschel, William. *The Scientific Papers,* edited (and with a biographical note) by J. L. E. Dreyer (London: Royal Society and Royal Astronomical Society, 1912).

Hoskin, Michael A. *William Herschel, Pioneer of Sidereal Astronomy* (London: Sheed and Ward, 1959).

Hoyt, William Graves. *Planets X and Pluto* (Tucson: University of Arizona Press, 1980).

Hoyt, William Graves. "William Henry Pickering's Planetary Predictions and the Discovery of Pluto," *Isis* 67 (1976): 551–564.

Hubbard, William B., André Brahic, B. Sicardy, L.-R. Elicer, F. Roques, and F. Vilas. "Occultation Detection of a Neptunian Ring-like Arc," *Nature* 319 (February 20, 1986): 636–640.

Hughes, David W. "Pluto, Charon and Eclipses," *Nature* 327 (May 14, 1987): 102–103.

Hunt, Garry, ed. *Uranus and the Outer Planets* (Cambridge: Cambridge University Press, 1982).

Ingersoll, Andrew P., "Jupiter and Saturn," *Scientific American* 245 (December 1981): 90–108.

Ingersoll, Andrew P. "Uranus," *Scientific American* 255 (January 1987): 38–45.

Jaki, Stanley L. "The Early History of the Titius-Bode Law," *American Journal of Physics* 40 (July 1972): 1014–1023.

Jaki, Stanley L. "The Original Formulation of the Titius-Bode Law," *Journal of the History of Astronomy* 3 (June 1972): 136–138.

Jaki, Stanley L. "The Titius-Bode Law: A Strange Bicentenary," *Sky and Telescope* 43 (May 1972): 280–281.

Johnson, Torrence V., Robert Hamilton Brown, and Lawrence A. Soderblom. "The Moons of Uranus," *Scientific American* 256 (April 1987): 48–60.

Jones, Harold Spencer. *John Couch Adams and the Discovery of Neptune* (Cambridge: Cambridge University Press, 1947).

Kane, Van R., and Charles E. Kohlhase. "Voyager to Uranus and Neptune," *Astronomy* 11 (April 1983): 6–15.

King, Henry C. *The History of the Telescope* (Cambridge: Sky Publishing, 1955).

Kohlhase, Charles E. "Aiming for Neptune," *Astronomy* 15 (November 1987): 6–15.

Kohlhase, Charles E. "On Course for Neptune," *Astronomy* 14 (November 1986): 6–15.

Kohlhase, Charles E. "Voyager's Path of Discovery," *Astronomy* 14 (February 1986): 14–22.

Kowal, Charles T., and Stillman Drake. "Galileo's Observation of Neptune," *Nature* 287 (September 25, 1980): 311–313. Also, discussion: *Nature* 290 (March 12, 1981): 164–165.

Laeser, Richard P., William I. McLaughlin, and Donna M. Wolff. "Engineering *Voyager 2*'s Encounter with Uranus," *Scientific American* 255 (November 1986): 36–45.

Lissauer, Jack J. "Shepherding Model for Neptune's Arc Ring," *Nature* 318 (December 12, 1985): 544–545.

Lowell, A. Lawrence. *Biography of Percival Lowell* (New York: Macmillan, 1935).

Lowell, Percival. *Memoir on a Trans-Neptunian Planet* (Lynn, Mass.: Press of T. P. Nichols, 1915; Memoirs of the Lowell Observatory 1.1).

Lubbock, Constance A. *The Herschel Chronicle* (Cambridge: Cambridge University Press, 1933).

Lyttleton, Raymond A. "On the Possible Results of an Encounter of Pluto with the Neptunian System," *Monthly Notices* of the Royal Astronomical Society 97 (December 1936): 108–115.

Maddox, John. "Whose Rings around Neptune?" *Nature* 318 (December 12, 1985): 505.

Matese, John J., and Daniel P. Whitmire. "Planet X and the Origin of the Shower and Steady State Flux of Short-Period Comets," *Icarus* 65 (1986): 37–50.

Maunder, E. Walter. *The Royal Observatory Greenwich* (London: Religious Tract Society, 1900).

McDonough, Thomas R. *Space: The Next Twenty-Five Years* (New York: John Wiley & Sons, 1987).

McKinnon, William B. "On the Origin of Triton and Pluto," *Nature* 311 (September 27, 1984): 355–358.

Macpherson, Hector. *Makers of Astronomy* (Oxford: Clarendon Press, 1933).

Miner, Ellis D. "On to Neptune!" *Planetary Report* 6 (November/December 1986): 17–18.

Moore, Patrick. "The Naming of Pluto," *Sky & Telescope* 68 (November 1984): 400–401.

O'Meara, Stephen James. "A Visual History of Uranus," *Sky & Telescope* 70 (November 1985): 411–414.

Pioneering the Space Frontier (New York: Bantam Books, 1986). Report of the National Commission on Space.

Porco, Carolyn C. "*Voyager 2* and the Uranian Rings," *Planetary Report* 6 (November/December 1986): 11–13.

Putnam, Roger Lowell, and V. M. Slipher. "Searching Out Pluto—Lowell's Trans-Neptunian Planet X," *Scientific Monthly* 34 (January 1932): 5–21.

Raup, David M., and J. John Sepkoski, Jr. "Periodicity of Extinctions in the Geologic Past," *Proceedings of the National Academy of Sciences USA* 81 (February 1984): 801–805.

Rawlins, Dennis. "The Unslandering of Sloppy Pierre," *Astronomy* 9 (September 1981): 24, 26, 28.

Ronan, Colin A. *Astronomers Royal* (Garden City, New York: Doubleday, 1969).

Roseveare, N. T. *Mercury's Perihelion from Lc Verrier to Einstein* (Oxford: Clarendon Press, 1982).

Sagan, Carl. *Murmurs of Earth: The Voyager Interstellar Record* (New York: Random House, 1978).

Sidgwick, J. B. *William Herschel: Explorer of the Heavens* (London: Faber and Faber, 1953).

Smart, W. M. "John Couch Adams and the Discovery of Neptune," *Occasional Notes* of the Royal Astronomical Society 11 (August 1947).

Smith, B. A., L. A. Soderblom, and 38 others. "*Voyager 2* in the Uranian System: Imaging Science Results," *Science* 233 (July 4, 1986): 43–64.

Smoluchowski, Roman, John N. Bahcall, and Mildred S. Matthews, eds. *The Galaxy and the Solar System* (Tucson: University of Arizona Press, 1986).

Stone, E. C., and E. D. Miner. "The *Voyager 2* Encounter with the Uranian System," *Science* 233 (July 4, 1986): 39–43.

Thiel, Rudolf. *And There Was Light: The Discovery of the Universe*, trans. by Richard and Clara Winston (New York: Alfred A. Knopf, 1957).

Tholen, David J., Marc W. Buie, Richard P. Binzel, and Marian L. Frueh. "Improved Orbital and Physical Parameters for the Pluto-Charon System," *Science* 237 (July 31, 1987): 512–514.

Tombaugh, Clyde W. "The Discovery of Pluto: Some Generally Unknown Aspects of the Story," *Mercury* 15 (May/June 1986): 66–72.

Tombaugh, Clyde W. "The Discovery of Pluto: Some Generally Unknown Aspects of the Story, Part II," *Mercury* 15 (July/August 1986): 98–102.

Tombaugh, Clyde W., and Patrick Moore. *Out of the Darkness: The Planet Pluto* (with foreword by James W. Christy on the discovery of Charon) (Harrisburg, Pennsylvania: Stackpole Books, 1980).

Tombaugh, Clyde W. "Reminiscences of the Discovery of Pluto," *Sky & Telescope* 19 (March 1960): 264–270.

Tombaugh, Clyde W. "The Search for the Ninth Planet, Pluto," *Mercury* 8 (January/February 1979): 4–6. Reprinted from *Leaflets of the Astronomical Society of the Pacific* 209 (July 1946).

Turner, Herbert Hall. "Obituary of Johann Gottfried Galle," *Monthly Notices* of the Royal Astronomical Society 71 (February 1911): 279.

Washburn, Mark. *Distant Encounters: The Exploration of Jupiter and Saturn* (San Diego: Harcourt Brace Jovanovich, 1983).

Whitmire, Daniel P., and Albert A. Jackson IV. "Are Periodic Mass Extinctions Driven by a Distant Solar Companion?" *Nature* 308 (April 19, 1984): 713–715.

Whitmire, Daniel P., and John J. Matese. "Periodic Comet Showers and Planet X," *Nature* 313 (January 3, 1985): 36–38.

Wilford, John Noble. *The Riddle of the Dinosaur* (New York: Alfred A. Knopf, 1986).

Williams, J. G., and G. S. Benson. "Resonances in the Neptune-Pluto System," *Astronomical Journal* 76 (March 1971): 167–177.

Yamamoto, Issei. "Professor Yamamoto's Suggestion on the Origin of Pluto," *Bulletin* of the Kwasan Observatory, Kyoto Imperial University 3.288 (August 20, 1934.)

Glossary

air glow Light in the upper atmosphere caused by solar radiation exciting atmospheric gases.

albedo The fraction of light that an object reflects.

altitude In astronomy, an angular measurement of position from the horizon upward. The zenith (the point overhead) has an altitude of 90 degrees from a level horizon.

aperture The diameter of the primary lens in a refracting telescope or the primary mirror of a reflecting telescope that defines the instrument's light-gathering power.

aphelion The point in an object's orbit around the Sun at which it is farthest from the Sun. (At this point, the object is traveling at its slowest speed.)

apoapsis The point in its orbit at which a body is farthest from the center of attraction.

apsis One of the points in an object's orbit at which it is closest or farthest from the object it is orbiting. (Periapsis is the closest point; apoapsis is the farthest.)

asteroid A planetlike body in the solar system too small to be classified as a planet. Most asteroids (or minor planets) orbit the Sun between Mars and Jupiter in a region called the asteroid belt.

astronomical unit (AU) A unit of measure in astronomy approximately equal to the average distance between the Earth and Sun: 92.96 million miles (149.6 million kilometers). Uranus, for example, is 30.1 astronomical units from the Sun.

asymptote The limiting value toward which a curve may tend. (A spacecraft traveling out of the solar system with escape velocity will continue to be slowed by the Sun's gravity, but less and less as its distance increases. Its velocity decreases toward but never reaches an asymptote.)

aurora Light radiated in the upper atmosphere in the region of a planet's magnetic poles because the atoms and molecules there are struck by charged particles from the Sun diverted toward the magnetic poles by the planet's magnetic field.

azimuth An angular measurement of position along the horizon, usually starting from north and moving clockwise. East is azimuth 90 degrees.

binary star Two stars revolving around one another.

black hole An object so dense that its gravity prevents light from escaping.

Bode's Law A numerical sequence that gives approximate planet distances from the Sun in astronomical units. (It fails for Neptune and Pluto.) This approximation was discovered by Johann Daniel Titius and popularized by Johann Elert Bode.

brown dwarf A celestial object with a mass slightly too low to initiate and sustain a nuclear reaction at its core and thereby make it a star.

charge-coupled device (CCD) A type of silicon wafer designed for the detection and measurement of light.

clathrate A molecular compound in solid form (such as water) in which molecules of another compound (such as methane) are trapped, but not chemically bound, in its cavities or cagelike crystals.

comet An object in orbit around the Sun composed of dust embedded in frozen gases.

conjunction (1) When a planet is in line with the Sun as seen from Earth; (2) when two or more planets appear to be very close together in the sky.

constellation A region of stars given a name to aid in remembering that star field.

dayglow Fluorescence of hydrogen atoms in the upper atmosphere on the sunlit side of a planet. This ultraviolet glow was discovered by the *Voyager* spacecraft at Jupiter, Saturn, and Uranus. Its cause is uncertain.

declination The angular distance of a celestial object measured north or south of the celestial equator. (Declination in astronomy corresponds to latitude in geography.)

density The mass of an object divided by its volume.

differentiation In an astronomical body, the separation of different kinds of materials from their original mixed state into different layers.

direct motion The gradual eastward apparent motion of a planet as seen from Earth.

Doppler Effect   The apparent change in wavelength of radiation (or sound) due to relative motion between the source and the observer along the line of sight. If source and observer are receding from one another, the wavelength is lengthened (redshifted). If the source and observer are approaching one another, the wavelength is shortened (blueshifted).

eccentricity (of an orbit)   A measure of the elongation of an ellipse, defined as the distance between the foci divided by the major axis.

ecliptic   The plane of the Earth's orbit around the Sun, hence also the apparent path of the Sun through the star field (constellations of the zodiac) in the course of one year as the Earth revolves around the Sun.

ellipse   A closed geometrical figure formed when a plane (flat surface) cuts through a cone at an angle but does not cut through the base of the cone. Planets, asteroids, and comets orbit the Sun in ellipses.

electroglow   (See dayglow.)

electromagnetic radiation   Radiation emitted when an electron falls to a lower energy level within an atom or molecule. Electromagnetic radiation travels at the speed of light and may be of any wavelength. From shortest wavelength (highest energy; highest frequency) to longest wavelength (lowest energy; lowest frequency), electromagnetic radiation includes: gamma rays, X-rays, ultraviolet light, visible light, infrared light (heat), and radio waves.

ephemeris (plural, *ephemerides*)   A listing that gives a celestial object's calculated positions for various times.

equatorial mount   A mounting for a telescope in which one axis points toward the celestial pole and the other axis is perpendicular to it so that a star or planet can be tracked by rotating only one axis to compensate for the rotation of the Earth.

fluorescence   The process in which a gas absorbs and then emits light. The emission of electromagnetic radiation (usually visible light but also gamma rays, X-rays, ultraviolet light, infrared light, or radio waves) by a substance that has absorbed energy of a shorter (higher-energy) wavelength.

Galilean satellites   The four brightest (and largest) satellites of Jupiter: Io, Europa, Ganymede, and Callisto. Galileo discovered these satellites in 1610.

giant planets   Jupiter, Saturn, Uranus, and Neptune.

graben   A depressed segment of crust bounded on at least two sides by faults and generally much longer than it is wide.

grand tour   In astronautics, a spaceflight that uses gravity assists to reach more than two bodies.

gravitation   The attraction of matter for other matter due to the mass possessed.

gravitational assist   The use of a massive moving body (such as a planet) to change the speed and direction of a passing spacecraft to reach an objective with a minimum expenditure of energy. (*Voyager 2* received gravitational assists from Jupiter, Saturn, Uranus, and Neptune.) Gravity assists also occur naturally. Comets sometimes receive gravity assists from Jupiter and are accelerated out of the solar system.

heliopause   An envelope around the Sun within which the Sun's magnetic field is dominant and beyond which interstellar fields dominate.

hydrocarbon compounds   Organic chemical compounds consisting of only hydrogen and carbon. (Methane is the simplest hydrocarbon.)

inclination of equator   The tilt of the equator of an astronomical body with respect to the plane of its orbit.

inclination of orbit   The tilt of the orbit of an astronomical body with respect to the orbital plane of another body.

infrared radiation   Electromagnetic radiation with wavelengths longer than visible light and shorter than radio waves. Infrared light is thermal radiation (heat).

invariable plane   The mean plane of the solar system. (It is inclined about 1.6 degrees to the ecliptic.)

ionize   To make an atom electrically charged by the loss or gain of one or more electrons.

Jovian   Relating to the planet Jupiter.

Kepler's laws   The three laws discovered by Johannes Kepler that describe planetary motion and the motion of any celestial body around another. The laws (expressed for planets) are, in very general terms: (1) Planets travel around the Sun in ellipses; (2) when a planet is near the Sun, it travels faster than when it is far from the Sun; (3) the larger a planet's orbit around the Sun, the slower the planet travels and the longer its revolution takes.

Lagrangian Points   When two bodies are revolving around each other in circular orbits, there are five points that lie in the plane of the revolving objects where a third body of negligible mass can remain in equilibrium with respect to the other two bodies.

latitude, celestial   The angular distance of a celestial object measured north or south of the ecliptic.

least-energy trajectory   In astronautics, a flight path from one location to another that is a portion of an ellipse with the destination at one apsis and the departure point at the other. Such a trajectory requires a minimum of energy (propellants) to accomplish the journey. (Same as orbital transfer trajectory.)

light-year A measure of distance in astronomy equal to the distance light travels in a year (approximately 5.88 trillion miles; 9.46 trillion kilometers; 63,240 astronomical units).

longitude, celestial The angular distance of a celestial object measured eastward from the vernal equinox along the ecliptic.

magnetometer An instrument for measuring magnetic intensity.

magnetosphere The region around a celestial body where its magnetic field dominates over the magnetic fields of surrounding bodies.

magnitude A measure of brightness in astronomy. The lower the magnitude number, the brighter the object. A star with apparent magnitude +1.0 is approximately two and a half times brighter than a star with apparent magnitude +2.0.

mean Average (as in mean plane of the solar system or mean distance).

meridian A giant circle in the sky that passes through the north and south points and the observer's zenith so that it divides the sky into eastern and western halves.

minor planet A planetlike body in the solar system too small to be classified as a planet. Most minor planets (or asteroids) orbit the Sun between Mars and Jupiter.

nebula A celestial cloud of gas and dust.

neutrino An elementary subatomic particle with no electrical change and no (or very little) mass.

node A point along the orbit of an object where it crosses a reference plane. In the case of a planet, its *ascending* node is where it crosses the plane of the Earth's orbit (ecliptic) going north. Its *descending* node is where it crosses the plane of the Earth's orbit going south.

occultation The blocking of the light of a celestial body by another celestial body usually of much larger apparent size; an eclipse of a star or moon by a planet.

Oort Cloud or Öpik-Oort Cloud A spherical swarm of trillions of comets surrounding our solar system. The Oort Cloud is now thought to be composed of a primordial inner disk beyond Pluto that lies in the plane of the solar system and an outer spherical cloud of comets scattered from the inner disk.

opposition When a planet is on the opposite side of the Earth from the Sun.

orbital transfer trajectory In astronautics, a flight path from one location to another that is a portion of an ellipse with the destination at one apsis and the departure point at the other. Such a trajectory requires a minimum of energy (propellants) to accomplish the journey. (Same as least-energy trajectory.)

organic chemical A chemical compound in which carbon is present.

periapsis The point in its orbit at which a body is closest to the center of attraction.

perihelion The point in an object's orbit around the Sun at which it is closest to the Sun. (At this point, it is traveling at its fastest speed.)

perturbation Deviation from the expected orbital motion of a celestial body.

photometer An instrument that measures light intensity.

photometry The measurement of light intensity.

photopolarimeter An instrument that measures light intensity and the amount that light is polarized. Polarization is a condition in light whereby the transverse vibrations of the rays assume different forms in different planes.

planetesimal One of numerous small solid bodies that existed at an early stage of solar system development from which the planets, moons, and asteroids formed. Comets are thought to be representative of the icy planetesimals that existed in the outer solar system.

planetoid Another name for a minor planet (or asteroid).

polymerization A chemical reaction in which small molecules combine to form larger molecules of repeating structural units.

precession, orbital The slow rotation of the axis of an orbit in the plane of that orbit.

quasar An astronomical object with an extremely large redshift, interpreted as a very distant object that is pouring out vast quantities of energy. It is thought that quasars are the nuclei of young galaxies. The name *quasar* is an abbreviation of its appearance to optical or radio telescopes as a quasi-stellar object or a quasi-stellar radio source.

reflecting telescope (reflector) A telescope that uses a mirror to gather light and direct it to a focus.

refracting telescope (refractor) A telescope that uses a lens to gather light and direct it to a focus.

residuals The differences between the results obtained by observation and by computation from a formula.

resolution In astronomy, the ability of an optical system to distinguish two objects or features on an object from a distance.

resonant orbits Orbits where objects revolving around the same center of attraction have periods that are in simple integer number relationship to one another (1 to 2; 2 to 3) so that the bodies always pass one another at the same points along their orbits.

retrograde motion   The apparent backward (westward) motion of a planet in the star field as seen from Earth due to the Earth catching up with and passing that planet.

retrograde revolution   The revolution of an object around the Sun in a direction opposite to the planets, or the revolution of an object around a planet in a direction opposite to the planet's rotation.

retrograde rotation   The rotation (spin) of an object on its axis opposite to its orbital motion.

revolution   The motion of one object around another. The Earth revolves around the Sun in one year.

right ascension   The angular distance of an object measured eastward from the vernal equinox along the celestial equator. (Right ascension in astronomy corresponds to longitude in geography.)

Roche Limit   The distance from the center of an astronomical body within which that body's tidal forces will prevent the formation of a satellite and will tear apart a large satellite that ventures within that zone.

rotation   The spin of a body on its axis. The Earth rotates once every 24 hours.

scarp   A line of cliffs produced by faulting or erosion. A fault scarp is a cliff or escarpment directly resulting from an uplift along one side of a fault.

seeing   The steadiness of the Earth's atmosphere as it affects the resolution that can be obtained in astronomical observations.

selenography   The science of the physical features of the Moon. (From Selēnē, Greek goddess of the Moon.)

short-period comet   A comet with an orbital period of less than 200 years.

solar wind   Fast-moving charged subatomic particles flowing outward from the Sun.

spectrogram   A photograph of a spectrum.

spectrograph   An instrument for photographing a spectrum.

spectrometer   An instrument for electronically measuring a spectrum.

spectroscope   An instrument for viewing a spectrum.

spectroscopy   The study of spectra.

spectrum   The array of colors or wavelengths obtained when light from an object is dispersed as it passes through a prism or diffraction grating. By examin-ing a spectrum, scientists can tell the temperature, chemical composition, radial velocity, spin, and many other things about a light source and the material that lies between the source and the observer.

speculum metal   A shiny metal alloy used to make the mirrors of reflecting telescopes. Speculum metal was usually copper and tin, sometimes with a little arsenic, antimony, or zinc added to improve the whiteness. This type of mirror was abandoned when it became possible to coat glass mirrors with a reflective surface.

sublimation   In chemistry, the passage directly from a solid to a gas without passing through a liquid state. (Frozen carbon dioxide [dry ice] sublimes at room temperature. Under the extremely low pressure conditions of space, water sublimes also.)

tectonics   Relating to the deformation of a planetary surface, especially folding, faulting, and crustal plate movement, due to internal forces.

tidal coupling   A gravitational effect in a two-body orbiting system in which one body is compelled by the gravity of the other to rotate and revolve in the same period of time, with the result that it always keeps the same face pointed toward the body it orbits.

Titius-Bode Rule   A numerical sequence that gives approximate planet distances from the Sun in astronomical units. (It fails for Neptune and Pluto.) This approximation was discovered by Johann Daniel Titius and popularized by Johann Elert Bode.

transit   (1) The passage of a small body (such as a moon) across the face of a larger body (such as a planet); (2) the passage of a celestial object across the meridian. (Timings of these events are used to pinpoint the positions of stars and planets.)

ultraviolet radiation   Electromagnetic radiation with wavelengths shorter than visible light but longer than X-rays.

zenith   The point in the sky directly overhead for an observer.

zodiac   A belt 18 degrees wide centered on the ecliptic that encircles the star field and encompasses the paths of all the planets except Pluto. The twelve constellations of the zodiac were especially famous despite the modest brightness of some because this zone was the only region visited by the Sun, Moon, and planets.

Index